High-Tech Europe

STUDIES IN INTERNATIONAL POLITICAL ECONOMY

Stephen D. Krasner and Miles Kahler, General Editors
Ernst B. Haas, Consulting Editor

High-Tech Europe

*The Politics
of International
Cooperation*

Wayne Sandholtz

UNIVERSITY OF CALIFORNIA PRESS
Berkeley · Los Angeles · Oxford

University of California Press
Berkeley and Los Angeles, California

University of California Press, Ltd.
Oxford, England

Library of Congress Cataloging-in-Publication Data

Sandholtz, Wayne.
    High-Tech Europe : the politics of international cooperation /
Wayne Sandholtz.
        p.   cm.—(Studies in international political economy)
    Includes bibliographical references and index.
    ISBN 978-0-520-30210-5 (alk. : paper)
    1. High technology industries—Europe—International cooperation.
2. Telecommunication equipment industry—Europe—International
cooperation. 3. Information technology—Europe—International
cooperation. 4. Telecommunication—Europe—International
cooperation. 5. Technology—Europe—International cooperation.
I. Title. II. Series.
HC240.9.H53S26 1992
338.4'762'00094—dc20   91-3633
                                                                    CIP

For Willis Arthur Sandholtz
and LaMyrl Boyack Sandholtz,
who encouraged me to wonder

Contents

List of Tables

Abbreviations

The following list contains most of the abbreviations and acronyms used in the text. It does not include those that are well known to North American readers (for example, IBM), those employed only once or twice and spelled out in the text, and those for which the abbreviation constitutes the actual name (for instance, CSF).

ACARD	Advisory Council on the Application of Research and Development (U.K.)
AI	artificial intelligence
AIP	advanced information processing
ANVAR	Agence Nationale pour la Valorisation de la Recherche
ASIC	application-specific integrated circuit
BECU	billion European Currency Units
BMFT	Bundesministerium für Forschung und Technologie
BOC	Bell operating company
BPO	British Post Office
BT	British Telecom
CAD	computer-aided design

CAM	computer-aided manufacturing
CCITT	International Consultative Committee for Telephones and Telegraph
CDC	Control Data Corporation
CEC	Commission of the European Communities
CEPII	Centre d'Etudes Prospectives et d'Informations Internationales
CEPT	Conférence Européenne des Administrations des Postes et Télécommunications
CERN	Centre Européen pour la Recherche Nucléaire
CESTA	Centre d'Etudes des Systèmes et des Technologies Avancées
CETS	Conférence Européenne de Télécommunications par Satellites
CII	Compagnie Internationale d'Informatique
CII-HB	Compagnie Internationale d'Informatique-Honeywell-Bull
CIM	computer-integrated manufacturing
CNCL	Commission Nationale pour les Communications et les Libertés
CNES	Centre Nationale d'Etudes Spatiales
CNR	Consiglio Nazionale delle Ricerche
COREPER	Comité des Representatives Permanents
COST	Coopération Scientifique et Technologique
CPE	customer-premises equipment
CPEMT	Comité de Politique Economique à Moyen Terme
CREST	Comité de Recherche Scientifique et Technique
CSA	Conseil Supérieur de l'Audiovisuel
DEC	Digital Equipment Corporation

DG XIII	Directorate General XIII for Telecommunications, Information Industries, and Innovation
DGRST	Délégation Générale à la Recherche Scientifique et Technique
DGT	Direction Générale des Télécommunications
DRAM	dynamic random-access memory
DTI	Department of Trade and Industry (U.K.)
EAB	ESPRIT Advisory Board
EC	European Community
ECLA	Economic Commission for Latin America
ECU	European Currency Unit
EEC	European Economic Community
EFTA	European Free Trade Association
ELDO	European Launcher Development Organization
EMC	ESPRIT Management Committee
EMU	economic and monetary union
ESA	European Space Agency
ESPRIT	European Strategic Programme for Research and Development in Information Technology
ESRO	European Space Research Organization
ETSI	European Telecommunications Standards Institute
EUREKA	European Research Coordination Agency
FAST	Forecasting and Assessment in the Field of Science and Technology (EC working group)
FCC	Federal Communications Commission (U.S.)
GATT	General Agreement on Tariffs and Trade

GDP	gross domestic product
GEC	General Electric Company, Limited
GIE	*groupement d'intérêt économique*
GNP	gross national product
HDTV	high-definition television
IBC	integrated broadband communications
IC	integrated circuit
ICL	International Computers Limited
ICT	International Computing Technologies
IDA	Integrated Digital Access
IMEC	Interuniversity MicroElectronic Center (Belgium)
INTUG	International Telecommunications Users Group
IO	international organization
IRI	Istituto per la Ricostruzione Industriale
ISDN	Integrated Services Digital Network
ISO	International Standards Organization
IT	information technologies
ITTF	Information Technologies Task Force
JESSI	Joint European Semiconductor Silicon project
LAREA/CEREM	Laboratoire de Recherche en Economie Apliquée/Centre d'Etudes et de Recherches sur l'Entreprise Multinationale
LLTRD	long lead-time R&D
LOP	*Loi d'orientation et programmation*
LSI	large-scale integration
MAP	Microprocessor Applications Project (U.K.)
MBB	Messerschmitt-Bölkow-Blohm

MDD	McDonnell Douglas
MECU	million European Currency Units
MISP	Microelectronics Industry Support Plan (U.K.)
MITI	Ministry for International Trade and Industry (Japan)
NASA	National Aeronautics and Space Administration (U.S.)
NEC	Nippon Electric Corporation
NET	Normes Européennes des Télécommunications
NIEO	New International Economic Order
NTT	Nippon Telegraph and Telephone
OECD	Organization for Economic Cooperation and Development
ONP	open-network provision
OPEC	Organization of Petroleum Exporting Countries
OSI	Open Systems Interconnection
OTA	Office of Technology Assessment (U.S.)
PABX	private automated branch exchange
PAFE	*Programme d'action pour la filière électronique*
PET	Planning Exercise in Telecommunications
PREST	Politique de Recherche Scientifique et Technologique
PTT	national administration of posts, telephone, and telegraph
RACE	R&D in Advanced Communications-Technologies in Europe
RAM	random-access memory
R&D	research and development

RTT	Régie des Téléphones et Télégraphes (Belgium)
SCI	Strategic Computing Initiative (U.S.)
SDI	Strategic Defense Initiative (U.S.)
SFI	Support for Innovation program (U.K.)
SGS	SGS-Ates Componenti Electtronici SpA
SIP	Societa Italiana per l'Esercizio Telefonico
SME	small and medium enterprises
SNA	Systems Network Architecture
SOGT	Senior Officials Group for Telecommunications
SPAG	Standards Promotion and Application Group
SRAM	static random-access memory
SRC	Science Research Council (U.K.)
STC	Standard Telephone and Cable (U.K.)
STET	Societa Finanziaria Telefonica
TIP	Technology Integration Project
UA	unit of account
UNICE	Union des Industries de la Communauté Européenne
VAN	value-added network
VHSIC	very-high-speed integrated circuit
VLSI	very-large-scale integration
WEU	Western European Union

Acknowledgments

A few sentences at the beginning of this book are grossly inadequate recompense to those who have generously aided and abetted me in this endeavor. Perhaps one day the flow of favors and kindnesses will reach a balance. At any rate I shall at least recognize my debts. Those ultimately responsible are my parents, Willis A. Sandholtz and LaMyrl Boyack Sandholtz, who encouraged my inquisitiveness and to whom this book is dedicated. I am grateful to Berkeley, the place and its people; I cannot imagine a more congenial and refreshing place in which to have completed this work.

Berkeley was full of superb colleagues who also happened to be friends. First mention must be of Ernst B. Haas, who was unstinting in sharing his time and insights and in providing the proper blend of hard-nosed criticism and warm encouragement. John Zysman and the rest of the crowd at the Berkeley Roundtable on the International Economy (BRIE) supplied abundant and good-natured support.

For financing this project, I thank the Institute of International Studies at Berkeley and the MacArthur Foundation (for grants received through the University of California at Berkeley and BRIE). The Institut Français des Relations Internationales (IFRI) supplied a friendly place in which to work while conducting field research in Europe; the staff at IFRI helped in invaluable ways, even in the smallest details. A Stevenson Faculty Development Grant from Scripps

College supported me during the summer of 1990 as I made final revisions.

I am profoundly grateful to the many in Europe who agreed to be interviewed for this project. Without them, the research would have been impossible. Though too many to name, my respondents generously contributed their time and intelligence, sometimes having to tolerate my uninspiring questions.

For reading and commenting on previous drafts (qualifying them for hardship pay), I thank Vinod Aggarwal, Ernst B. Haas, Jeffrey Hart, Thomas Ilgen, Susan Sell, Kenneth Waltz, Wes Young, and John Zysman. Robert Switky provided able research assistance in the final stages of preparing the manuscript. Above all, I thank my wife, Judy Haymore Sandholtz, the perfect partner.

Wayne Sandholtz
Claremont, California

ONE

Introduction

The Puzzle of European High-Tech Cooperation

Technology and the state long ago forged bonds of mutual dependence. Each satisfies profound cravings of the other. The state nourishes technology with cash, making possible the grandiose projects that would otherwise be lustful dreams of technologists and engineers. By the same token, technology provides the state with industries, wealth, and tools of defense. In the late twentieth century, though, the relation between technology and the state appears unbalanced: Technology can survive without the modern state, but the reverse may not be true.

The harnessing of technology to purposes dictated by nationalism is as old as the manufactories and arsenals of the European monarchies. But, coupled with modern science, technology has reached previously unimaginable degrees of scale and organization in the past several decades. Political leaders discovered that state-supported science and technology could create new sources of military power and decisively alter the terms of international politics. The most dramatic demonstrations emerged under the pressure of war: rockets, radar, and, ultimately, the atomic bomb. But, equally important, science and technology became the recognized basis for economic development and national prosperity. Governments in the industrialized countries saw that high technology could alter comparative advantage and generate new sources of wealth. Thus, nations compete technologically, in both security and economic realms, and nationalism has colored efforts to nurture and exploit high technology.

When Britain, France, Germany, and their neighbors agree to the joint development of crucial technologies, therefore, their cooperation demands a close look. The countries of Western Europe initiated in the 1980s a series of collaborative programs for research and development (R&D) in forefront technologies. The collaboration embodied in programs like the European Strategic Programme for Research and Development in Information Technology (ESPRIT), R&D in Advanced Communications-Technologies in Europe (RACE), and the European Research Coordination Agency (EUREKA) centers on the crucial telematics technologies—semiconductors, computing, and communications. Such collaborative efforts present us with a puzzle: Why, and how, did the Western European countries decide to cooperate in telematics, an area dominated by nationalism and competition? This book addresses that puzzle, examining the process by which states establish cooperation. The analytical focus will be on policy adaptation, political leadership, and international institutions.

The puzzle of telematics collaboration in Europe embraces a number of phenomena to challenge students of international politics. The problems posed by ESPRIT, RACE, and EUREKA will also interest students of technology and industrial policies. First, why are the countries of Western Europe collaborating in technologies of such singular economic and military importance? Telematics technologies are transforming every segment of the economy, from manufacturing to banking to retailing, health care, and entertainment. The same technologies provide the "brains" for modern weapons systems. Traditional realist theory would consider the telematics sectors to be among the least likely candidates for collaboration, given their importance for national power and relative position in the world. The simple structural realist explanation would be that the Europeans cooperate because they are weak. But that explanation is exactly as uninteresting as it is true. It verges on the tautological; naturally, if they were not weak, they would not cooperate. Furthermore, the weak do not inevitably collaborate. Other factors must weigh in the explanation.

A second question is: Why are the Europeans cooperating now when they did not do so before? European deficiencies in telematics were obvious in the 1960s—twenty years before collaboration emerged. If weakness explains collaboration, why did it not begin before 1980? Technology gaps became a cause célèbre in Europe in the 1960s, but collaboration did not follow. Indeed, for twenty years, attempts to initiate collaboration repeatedly foundered. In the 1980s ESPRIT, RACE, and EUREKA broke the national monopolies on telematics policy-making. Between the previous crisis and the 1980s how had the technologies, the competitive environment, and the perceptions of policy-makers changed?

Third, why do the collaborative programs not limit themselves to basic scientific research? Far removed from competing commercial interests, cooperation in basic research is less challenging politically than it is in applied research. In fact, the prevailing hypothesis has been that scientific cooperation is easy—as in the Centre Européen pour la Recherche Nucléaire (CERN), with its Nobel Prizes—but when it comes to technologies with commercial potential, national rivalries take over and scuttle attempts at cooperation. Euratom furnishes a pertinent example. ESPRIT and the other programs studied here contradict that argument by encompassing technologies with important commercial implications in the short and medium terms. Furthermore, the programs are not simply a handful of small, ad hoc projects. Two of them—ESPRIT and RACE—have clear strategies and detailed work programs. All three have significant budgets: ESPRIT with a total of 4.7 BECU; RACE with 1.1 BECU; and EUREKA with nearly 6 BECU.[1]

THE EMPIRICAL TERRITORY

Three major collaborative R&D programs came into existence between 1982 and 1985 in Western Europe. Two programs sponsored by the European Communities (EC), ESPRIT and RACE, take aim at Europe's deficiencies in information technologies and telecommunications.[2] A third, EUREKA, constituted and managed outside

1. The European Currency Unit (ECU) was worth 0.79 U.S. dollars in 1984 and 0.85 U.S. dollars in 1986. In early 1990 the ECU was trading at approximately 1.20 U.S. dollars. MECU means million ECU, and BECU means billion ECU.

2. A third EC program, Basic Research in Industrial Technologies for Europe (BRITE), deals with manufacturing technologies. I do not include BRITE in this study because it does not have the financial scale nor the visibility of ESPRIT and RACE, and because ESPRIT and RACE provide more than enough insight into EC R&D efforts. In other words, including BRITE would not appreciably increase the analytical reach of this study. Also, by limiting the cases to programs concerned solely or primarily with telematics, I restrict the number of technologies involved and therefore keep the number of technology-related variables manageable.

EC structures, includes other technological areas (like biotechnology, new materials, and the environment). Even so, the largest share (30 percent) of EUREKA projects deals with telematics. Thumbnail sketches of these programs follow.

ESPRIT

Serious concerns for the health of Europe's telematics industries coalesced within the Commission of the European Communities (CEC) in the late 1970s. A new commissioner for industry, Etienne Davignon, joined the Commission in 1979 and lent his encouragement and bureaucratic protection to a small group within the Commission that was studying Europe's needs in the information technologies (or IT, which includes semiconductors and computers and their applications). Davignon and his associates were convinced that Europe's continued economic growth depended on strength in IT, and that Europe's IT industries were falling dangerously behind their American and Japanese competitors. In late 1981 Davignon invited the directors of the twelve largest European IT companies to round-table discussions on the future of the industry in Europe. These meetings were followed by technical discussions among the companies. By early 1982 the twelve were able to present a work plan for a collaborative R&D program, specifying the technologies and objectives to be pursued. A year later the ESPRIT pilot phase was launched, attracting over two hundred proposals (only thirty-eight could be funded). In February 1984 the Council of Research Ministers approved Phase I of the ESPRIT program. It carried a budget of 1.5 BECU, half of that amount coming from the EC and half from project participants. Groupings of EC companies, universities, and public laboratories submit proposals, and experts retained by the Commission select the best from among them for funding.

So enthusiastic was the response from the IT community that the Phase I monies, originally planned to last five years, were exhausted in three. The Commission asked the Council to move up the starting of ESPRIT Phase II from 1989 to 1987. Phase II became caught in the logjam surrounding the Framework Programme (the EC's overall R&D plan) but was approved in autumn 1987 at about 3.2 BECU (half from the EC) over five years.

RACE

RACE was in some ways a spinoff from ESPRIT. The notion of a collaborative R&D program in telecommunications originated with the same Commission task force that was behind ESPRIT. In addition, the same twelve companies provided the technologists who prepared a detailed plan for collaboration in telecommunications R&D, standards setting, and planning the next-generation network in Europe. This plan served as the basis for the Commission's proposal for a RACE Definition Phase, submitted to the Council in March 1985 and approved four months later.

The Commission's task in RACE, however, was complicated by the traditional monopoly on telecommunications networks and services held by the national telecoms administrations (the PTTs). The PTTs insisted that the work of defining a future broadband network take place in the Conférence Européenne des Administrations des Postes et Télécommunications (CEPT), the association of PTTs. In the end a compromise provided for close cooperation between the CEPT and the RACE program in preparing a reference model for the future broadband network and in setting standards.

The full RACE program became stuck in the wrangling over the Framework Programme in late 1986 and early 1987. However, those states holding up the Framework Programme (chiefly Germany and Great Britain) objected to other parts of the program and not specifically to RACE. The RACE Main Phase was approved in the fall of 1987 with an EC contribution of 550 MECU, to be matched by the participants. The program operates on the same basis as ESPRIT: Consortia of companies, universities, and public laboratories from the member countries submit proposals to the Commission, which selects projects with the help of outside experts.

EUREKA

EUREKA arrived on the scene via a completely different route, though it responded to the same fears about the status of high technology that motivated ESPRIT and RACE. French President François Mitterrand proposed a joint R&D program to counteract the civilian repercussions of the U.S. Strategic Defense Initiative (SDI). The French argued (and many Europeans agreed) that SDI would harm Europe's high-tech enterprises by (1) subsidizing technological advances in U.S. industries that would translate into market advantages; (2) inducing a brain drain, as European scientists and technologists followed the allure of cash-rich SDI projects; and (3) inhibiting the flow of resulting breakthroughs to Europe by imposing security controls on technology transfer.

In proposing EUREKA President Mitterrand was acting on his desire to take the lead in building Europe's high-tech community. Thus, the EUREKA initiative was a response to SDI but only because the American program touched on an already sensitive nerve in Europe. Worth noting is that although President Mitterrand's proposal in April 1985 contained no details as to structure or content for the program, EUREKA was approved in general outline by seventeen European states in mid-July, a mere blink of the eyes in the history of European cooperation. EUREKA took on shape and substance over the next year. The participating countries (including several non-EC states) gave the program a seven-person secretariat outside of EC institutions. The secretariat acts as a clearinghouse, granting no funds of its own. Consortia of companies, universities, and laboratories from at least two countries submit projects, and participants request financial assistance from their own national authorities. By the end of 1986 projects worth over 3.5 BECU had received the EUREKA label. The total value of announced EUREKA projects (through the Vienna meeting of EUREKA ministers) was almost 6 BECU in 1990.

ESPRIT, RACE, and EUREKA are at the heart of the upsurge in technological collaboration that has been taking place in Western Europe since the early 1980s. Exploring these three cases will reveal links among international collaboration, change in government policy preferences, private transnational actors, technological change, and international organizations. ESPRIT demonstrates both the vital part played by a transnational coalition of major companies and the leadership potential of an international organization. RACE highlights the role of national bureaucracies (the PTTs) in an area where international cooperation is emerging. The EUREKA case provides a useful counterpoint to the others because it took shape outside EC institutions. I will also draw on other instances of technological collaboration in Europe (like Airbus and the European Space Agency) for comparative perspectives on some points.

THE ANALYTICAL APPROACH

For scholarship on international cooperation the joint technology programs in Western Europe during the 1980s provide fresh empirical materials. In this section I briefly lay out my approach to theorizing about international cooperation, and in the next I introduce an analytical framework.

Phenomena that attract the attention of social scientists are invariably complex and multidimensional. The explanatory variables are frequently linked (something like autocorrelation in econometric models). Outcomes interact with the "independent" variables in feedback mechanisms. In many cases the historical context, the timing of events, and the beliefs of individual decision-makers produce phenomena that, in important respects, are unique. As a result strong causal relations are elusive; they are frequently so general as to be inapplicable or so qualified as to be useless.

Even when we have strong theoretical constructs, the task of empirical verification is not a straightforward one. Contending theories can seldom have their high noon in a crucial empirical test. Rival theories frequently stem from alternative views of the world and of the nature of social science inquiry; they embody distinct views of the goal of analysis. As a result there are possibly insurmountable obstacles to erecting decisive tests: Data do not exist except in the context of a theory, and the terms (the theoretical language) of one theory cannot be translated into those of another.[3] What, then, is a social scientist to do?

Fortunately, there are multiple paths to the temples of science. I follow Stephen Toulmin in arguing that the goal of science is understanding, and understanding enables the scholar to supply satisfying explanations of natural and social phenomena.[4] I have two goals in this book: to produce a satisfying explanation of telematics collaboration in Europe, and to propose and employ a general analytical approach for explaining international cooperation. In other words, I am not advancing a theory of international cooperation but rather a strategy for theorizing.

THE ARGUMENT IN BRIEF

The analysis begins with fundamental axioms of international-relations scholarship. Because no supranational authority exists to

---

3. See Paul Roth, *Meaning and Method in the Social Sciences*.
4. Stephen Toulmin, *Foresight and Understanding* and *Human Understanding*.

guarantee the security and independence of nation-states, each country must rely on its own means. For this reason leaders of states prefer autonomy to dependence or even to cooperation. State leaders quickly realize that in some areas cooperation is the only way to attain valued ends: There must be, for instance, international rules on traffic patterns for passenger aircraft. In addition states with ample resources are able to pursue autonomy in more areas and for more extended periods than smaller states can. Despite such qualifications, the generalized, built-in preference for autonomy makes international cooperation a phenomenon to be studied.

Any explanation of international cooperation, I propose, must address three key features of the phenomenon. The first two correspond roughly to demand and supply, though I use the terms heuristically, not as part of an economics-based model. Analysis must account for the demand for cooperation: Why do states choose to cooperate when their universal preference would be for autonomy? The question of demand is one, therefore, of beliefs and preferences. The scholar must be able to analyze how state leaders define preferences and how those preferences change.

In order for cooperation to emerge state leaders must become persuaded that unilateral means are insufficient to attain valued ends. The first task of analysis is therefore to explain the cognitive process through which governments come to recognize the inadequacy of unilateral approaches and begin to search for alternatives. I propose two broad kinds of cognitive change—learning and adaptation. Adaptation is relevant for the cases of ESPRIT, RACE, and EUREKA. Adaptation, or learning from experience, occurs when countries attempt unilateral strategies and discover that they are incapable of producing desired results. At this point policy-makers begin to search for new approaches. Cooperation is by no means the only (or even necessarily the most likely) outcome of that search. However, I hypothesize that the adaptive process is necessary for cooperation to occur. As long as unilateral measures might achieve the sought-after ends, states will not consider collaboration.

European technological collaboration seems to confirm this hypothesis. Airbus, for instance, did not coalesce until after independent French and British efforts had proven to be incapable of producing a commercially viable jetliner. The telematics programs of the 1980s also confirm this proposition. After the technology-gap crisis of the 1960s European governments went about busily trying to implement unilateral, national-champion strategies. They were not interested in collaboration in telematics until they realized that these national-champion strategies were inadequate. By 1980 unilateral policies had failed to close any of the gaps, and European governments were searching for new approaches. In that setting ESPRIT, RACE, and EUREKA could receive favorable consideration in national capitals. Governments were in the process of adaptation in response to policy failure.

The second step in explaining international cooperation is to account for the supply of political leadership. As Robert Axelrod has shown, cooperation can arise spontaneously out of the interactions among self-interested actors.[5] Though that may be the case in principle, cooperation in international politics is frequently difficult to achieve even when some states want it. As rational-choice theorists have taught us, collective action often depends on political leadership to organize it.

Political leaders propose collaboration and mobilize support for it. I argue that international political leadership is necessary for collaboration to arise. Leadership can come from one or a few key countries or, as I will argue, from an international organization. In the case of ESPRIT and RACE the Commission exercised the leadership role. The Commission first mobilized a transnational coalition in the telematics industry; then, with support from those influential firms, it won support for the programs from the national governments. With EUREKA France led.

The third and final element in this analytical framework is fair returns. Whether the effort to organize collaboration succeeds depends on whether states can agree on a cooperative arrangement that satisfies the interest calculations of each of them. I argue, contrary to some neorealists, that state leaders do not invariably base their decisions on calculations of relative gains or losses.[6] Rather, states seek a balance between their contributions to the cooperative effort and the benefits they receive from it—that is, states must find satisfactory solutions to the problem of *juste retour*. In Chapter 2 I suggest two distinct kinds of solutions. For the moment it is sufficient to point out that Airbus and ESPRIT embody different solutions to this problem. In Airbus the balance of contributions and

5. Robert Axelrod, *The Evolution of Cooperation.*
6. See Joseph M. Grieco, "Anarchy and the Limits of Cooperation: A Realist Critique of the Newest Liberal Institutionalism."

benefits is formally agreed on. In ESPRIT the work is divided into hundreds of projects of varying sizes, with participation in each project by self-selection. With such "à la carte" participation and a large number of projects, chances are greatly increased that self-selection will lead to a satisfactory share of the work. As will be seen in Chapter 5 the European Space Agency employs both these solutions simultaneously: contributions and benefits balanced by formal agreement, and à la carte participation.

In short I propose that at the most abstract level any explanation of international cooperation must include three elements. First, it must account for the cognitive process by which states adapt and settle for something other than unilateral autonomy—the question of demand. Second, the explanation must account for the supply of international political leadership. And, third, it must show how states resolve the problem of *juste retour*.

THE PLAN OF ATTACK

In Chapter 2 I lay out the analytical framework for discussing the politics of international cooperation. Readers not interested in theories of international politics can skip Chapter 2. Chapter 3 introduces the telematics technologies and portrays their economic importance in the industrialized societies. Chapter 4 summarizes the evolution of science and technology policies in the welfare democracies of Western Europe and describes the national-champion strategies that prevailed in the major countries up through 1980. The fifth chapter draws lessons from Europe's collaborative successes in aerospace and its failed attempts to cooperate in telematics during the 1970s. Chapter 6 depicts the 1980 crisis that confirmed the failure of national-champion policies. It describes what European policy-makers were seeing clearly: that unilateral strategies did not make European telematics industries sufficiently competitive with their American and Japanese rivals. It also analyzes the technological and economic changes that led the major telematics firms to favor collaboration. Chapters 7 through 9 analyze the three case studies, ESPRIT, RACE, and EUREKA. Finally, Chapter 10 reprises the analytical themes and compares the case studies with data from Euratom, Unidata, Airbus, and the European Space Agency.

TWO

The Politics
of International Cooperation

Though international politics seem at times to be a mélange of the irrational, the idiosyncratic, and the absurd, this study assumes that the main characters in the drama are rational actors. Leaders of states and international organizations choose among options so as to maximize the attainment of *their* values, not those assumed or assigned by scholars. Their rationality is bounded, and they choose in a context defined by both international and domestic politics and institutions. Beliefs, ideas, norms, and institutions matter in decision-making.

My analytical framework builds on these assumptions. It adds to the tradition in international-relations scholarship that is sometimes called neoliberal. Neoliberal analyses tend to explain international politics from the bottom up—that is, the international "system" is the sum of actor choices and behaviors. Regularities at the system level consist of patterns of behaviors. As Ernst Haas puts it, "The actors' perceptions of reality result in policies that shape events; these effects then create a new reality whose impact will then be perceived all over again."[1]

In contrast, the mainstream neorealist, or structural, tradition argues from the top down. The system, defined by its structure, sets limits to the choices of actors. The system is an independent vari-

1. Ernst B. Haas, "Words Can Hurt You; or, Who Said What to Whom about Regimes," 57.

able, not just the sum of patterned behaviors.[2] The national interest derives from a state's position in the system, and, as a result, relative power is the primary consideration driving the decisions of leaders. Thus, systemic constraints and relative power are seen as explaining everything from alliances to the New International Economic Order (NIEO).

THE CASE AGAINST STRUCTURAL EXPLANATIONS

Structural explanations do not provide satisfactory explanations of technological collaboration in Europe, much less of international cooperation more generally. Their utility is limited to the most obvious parts of the phenomenon; they leave a great deal unexplained. A structural explanation might start with the proposition that the decisive factor in the emergence of telematics collaboration in the 1980s was Europe's relative international weakness in those technologies. European firms were faring poorly in competition with U.S. and Japanese companies. To argue in a realist mode, the technologies were seen as important to national interests, and collaboration was a defensive alliance. At one level this explanation is true: European countries would not have looked to collaboration if they could have competed successfully on their own.

But that argument verges on the tautological: The weak cooperate because they are weak. Furthermore, there are many steps between weakness and collaboration that the structural logic leaves unexplored. The structural argument implies that there are no viable choices for the weak other than cooperation. I argue, in contrast, that even the weak have options, and structure falls far short of explaining why they select one option and not the others. Some scholars in the realist tradition recognize that nonstructural variables matter, though they assign them secondary importance. For instance, Stephen Krasner has argued that the NIEO was the response of the Third World to relative weakness in the international structure. But Krasner's explanation of the emergence of the Group of 77 and the NIEO also depends crucially on important nonstructural variables—namely, a set of ideas that Third World leaders could agree on and political leadership within the group.[3] The Third

2. The classic expression of the neorealist approach is that of Kenneth N. Waltz, *Theory of International Politics.*
3. Stephen D. Krasner, *Structural Conflict.*

World achieved a degree of unity because of a set of ideas that could win consensus and a set of leaders who could mobilize a coalition.

More generally, there are no logical links leading from international weakness to alliances with similar partners. There are always other options, even for the weak, and the system/structure approach cannot tell us why one alternative will be chosen over another. Sometimes small or weak states seek out large, powerful protectors. Sometimes they remain fiercely independent. With respect to European telematics collaboration, why did not each state pursue a renewed national-champion strategy? Furthermore, the system/ structure approach cannot account for the timing of collaboration. If weakness produces alliance, why did the European countries not collaborate fifteen years earlier, in the technology-gap crisis of the 1960s?

Perhaps, a realist might counter, the rise of Japan as a technological power constituted a structural change that precipitated the European alliance. True, in part: Japanese IT producers constituted formidable and dangerous new competitors for weak European firms. Even so, European policy-makers could choose among options. Japan could become an alternative to the United States as a source of forefront technologies, driving down the costs of catching up. By seeking ties with the Japanese, the Europeans could extract better deals from the Americans. Or state leaders in Europe could respond through revived national-champion strategies. Or they could supplement unilateral policies with selective bilateral arrangements, such as the Mega Project involving Philips, Siemens, and their governments.

Another kind of structural argument would focus on the structure of capabilities within the set of cooperating states. Charles Kindleberger explored this line of thinking. He posited that a liberal international economic system depends on the existence of a predominant state, or hegemon.[4] System/structure logic (in its hegemonic-stability form) would imply that a predominant power in Europe would be necessary to provide a stable collaborative arrangement. Against this argument some scholars have shown that, in principle, a small group of states, none of them predominant, with similar interests can initiate and sustain cooperation.[5] With

4. Charles Kindleberger, *The World in Depression.*
5. See Robert O. Keohane, *After Hegemony;* and Duncan Snidal, "The Limits of Hegemonic Stability Theory."

respect to the specific concerns of this study, ESPRIT, RACE, and EUREKA all took shape without the benefit of a dominant state in Europe. Britain, France, and Germany were approximate equals in the telematics industries, with Italy not far behind. In fact, Britain, France, and Germany each had three companies among Europe's twelve largest electronics firms. Italy had two, and the Netherlands, one formidable telematics player in Philips. Clearly, no hegemon held sway in Europe. This study thus provides an empirical case of cooperation emerging among states without a hegemon.

In short, there simply is no structural explanation of technological collaboration or of cooperation in general. No logical necessity connects any given configuration of the international system with any particular outcome. It is not enough to point out that the European states were weak in telematics relative to their international rivals. Of course, the Europeans would not have worried about their competitiveness if they had been competitive. Weakness in IT was a challenge to which more than one response was possible; system/ structure arguments cannot explain why cooperation should emerge rather than the alternatives. The international system can pose challenges and opportunities to national leaders, but it cannot explain their responses.

THE THEORETICAL FRAMEWORK

Joseph Nye assessed the common themes connecting neoliberal thinkers from the pioneering works of Ernst Haas, Karl Deutsch, and Nye to the present. All share

> a focus on the ways in which increased transactions and contacts changed attitudes and transnational coalition opportunities, and the ways in which institutions help foster such interaction. In short, they emphasized the political process of learning and of redefining national interests, as encouraged by institutional frameworks and regimes.[6]

Nye's statement of the common thread in neoliberal analyses embraces the present study. The key analytical theme in examining European telematics collaboration is precisely the "political process of learning and of redefining national interests," with a focus on how international institutions can stimulate and lead in that process.

6. Joseph S. Nye, Jr., "Neorealism and Neoliberalism," 239.

Nation-states must rely principally on their own means to ensure their security, survival, and prosperity. Leaders of nations seek to maximize their autonomy from other states. Assuming such a preference for autonomy does not require that one build on it the same theoretical superstructure that realists do. Rather, it seems more fruitful to think of the preference for autonomy as leading to a problem of collective action at the international level. Cooperation necessarily imposes constraints on national leaders, foreclosing some options, requiring others, or both. Cooperation implies some degree of policy adjustment. By this definition, when states pursue policies that are mutually beneficial and do not require any policy adjustments, that is harmony of interests, not cooperation.[7] State leaders in general, therefore, prefer autonomy to cooperation. Students of international politics face two major tasks. First, we must show why state leaders choose courses that limit their international autonomy; we must account for the demand for cooperation. Second, we must address the collective-action problem at the international level.

To explain the paths from a preference for autonomy to international cooperation, I propose a three-part theoretical framework. The first two parts deal with demand and supply: the demand for cooperation and the supply of political leadership. Robert Keohane employed the same terminology in explaining the demand for regimes.[8] Like Keohane, I use the notions of demand and supply heuristically, not to imply a market for international cooperation. My approach differs from his in two respects, though. First, I analyze the demand for cooperation not by how regimes meet functional needs of states but rather by how the preferences and beliefs of state leaders shift so as to make cooperation possible. Second, unlike Keohane, I devote equal attention to the supply of leadership.

The third stage of analysis is to examine the arrangements for distributing the costs and benefits of cooperation. The theoretical proposition involved is that state leaders must perceive an adequate balance between contributions and rewards, or they will not participate. With technological collaboration this question takes on the specific meaning of finding solutions to the problem of *juste retour*. I will propose two potential mechanisms for resolving that problem.

7. Keohane makes a similar argument in *After Hegemony*.
8. Robert O. Keohane, "The Demand for International Regimes."

What follows is not a general theory of international cooperation. I doubt that a single theory could adequately explain the various species of the phenomenon. The analytical framework developed below builds on existing theories and binds them into a whole
that is, I suggest, larger than the sum of its parts. It is a strategy
for analysis, a guide for structuring explanations.

Cognitive Change

The demand for cooperation involves cognitive change. For cooperation to emerge state leaders must come to believe that unilateral
means will not achieve valued ends. This explanation implies rationality. I assume that policy-makers and organizations can learn.
All the players in the drama of international cooperation (politicians, bureaucrats, officials of international organizations, industry
executives) are rational actors. But their rationality is the bounded
rationality described by Herbert Simon. They are rational in that
they choose among options so as to maximize the attainment of
values on the basis of available information and resources. They
are not capable of ranking all preferences, perceiving every alternative, foreseeing all consequences, or making all relevant value
tradeoffs.[9] Actor preferences are not deducible solely from bureaucratic position, electoral considerations, personal economic interests, or any other single, reductionist source. The sources of preferences are complex and include, in addition to the factors just
mentioned, ideologies and beliefs about causality (the relation between ends and means).

Two different kinds of cognitive change can lead policy-makers
to conclude that unilateral strategies are insufficient. Each cognitive
process corresponds to a different class of international problem.
For some kinds of problems it is clear that other states are inextricably involved in any solution—that is, solutions require policy
adjustments in more than one state; unilateral approaches possess
little or no value. In some instances, this need for coordination may
be obvious, as in the handling of international mail. Sometimes it
is not at all obvious that a problem is multilateral in nature. Indeed,
it may not even be clear that a problem exists. For instance, before
there was a technical understanding of the degradation of the ozone

9. Herbert A. Simon, *Administrative Behavior*.

layer, it was not the clear candidate for multilateral concern that it became with the 1987 Montreal Protocols to limit emissions of chlorofluorocarbons. I label this class of issues problems of scope because their resolution inherently requires the participation of more than one state.

Problems of scope like the handling of international mail are obvious and hence only marginally interesting. Other problems of scope require cognitive evolution on the part of national leaders. The type of cognitive change involved is *learning* as defined by Ernst Haas: the use of consensual knowledge to redefine causal relations so as to affect public policy. Learning entails the "sharing of larger meanings among those who learn." *Consensual knowledge* refers to beliefs about cause/effect linkages that are derived from "information, scientific and nonscientific, available about a given subject and considered authoritative by the interested parties." It is social knowledge, subject to evolution and revision.[10]

Scientists and experts generate new information and theories; sometimes these acquire broad agreement in the relevant scientific community. Of course, different political actors can use different bodies of scientific knowledge (or different interpretations of a single body of knowledge) to push for contrary policy options. So scientific facts and theories become consensual knowledge for public decision-making only when they are accepted as valid for policy-making purposes by the relevant politicians, interest groups, and bureaucrats. International cooperation can occur when national leaders reach a consensus on the relevant technical information. One way to apply the notion of learning is by focusing on epistemic communities, the networks of actors who develop and diffuse consensual ideas on policy problems.[11]

Problems of scale constitute a separate category. They do not inherently require multilateral solutions. The crucial concern with problems of scale is not how the problem transcends national boundaries but rather how to muster the level of resources needed to address the problem. Some states are large enough or wealthy enough to go it alone on many issues. Other states cannot marshal the resources needed at acceptable political cost. The United States has so far been able to sustain two companies producing civilian

10. Ernst B. Haas, *When Knowledge Is Power,* 21–24.
11. An interesting example of this approach is Peter M. Haas, "Do Regimes Matter? Epistemic Communities and Mediterranean Pollution Control."

airliners, Boeing and McDonnell Douglas. But four European countries found that they could not individually sustain aircraft industries and they banded together in Airbus.

A crucial difficulty is that there is no objective level of resources
needed to address any given problem of scale. Ambiguity clouds the
issue because so much depends on the political component, on how
many resources political authorities are able to devote to the problem. In problems of scale, the viability of unilateral solutions depends both on the scale of resources potentially available and on
the capacity of political actors to mobilize and focus those resources. For example, a small or poor country can autonomously
pursue large-scale projects if the political system is capable of focusing resources. India's space program may fit in this category.
Much depends on the value governments place on autonomy in a
given area and on the opportunity costs they are politically able to
accept in order to pursue it.

In other words, with problems of scale the resource threshold is
in part a matter of perception and political capacity. Furthermore,
national leaders may not know the cost of unilateral policies in advance, much less whether they can afford them. The outside observer cannot know either; countries with apparently limited resources can be quite dogged in pursuing autonomy. As a consequence,
I propose, state leaders frequently learn through experience the price
of autonomy and the political limits of their resources.[12] The Mitterrand government in France learned from experience in 1981–
1983 that it could not afford a unilateral policy of reflation. Because states prefer autonomy, I hypothesize that they will generally
attempt unilateral strategies first and surrender the goal of autonomy only when unilateral means have proven to be impossible or
too costly.

The process by which states discover the practical limits to autonomy is a species of cognitive change that I call adaptation. Adaptation is a less far-reaching cognitive shift than learning, as defined above. It entails a reconsideration of the means chosen to pursue
policy objectives. In adaptation decision-makers do not revalue their
ultimate ends in light of consensual knowledge; they search for new

12. Technically countries or states do nothing; people acting in or for them do.
When I write that states "choose" or "learn" something, I use the word *state* as
shorthand for the leaders of the state.

means to achieve those ends.[13] Of course, state leaders link objectives in hierarchies or chains, with lower-level goals serving as means to higher-level ones.[14] At the most abstract level (for example, "economic growth" as a national goal) these policy ends probably never change. A lower-level goal might be "national capacity in high-technology sectors," seen as necessary to achieve economic growth. Adaptation at this link in the ends/means chain is probably extremely gradual. Subordinate to high-tech capacity might be the goal of sustaining on national soil an enterprise capable of competing in global markets, a national champion. Adaptation is probably more common at this lower level than at higher levels. The adaptations examined in this study take place among subordinate means. Not even the mid-level goal of high-tech capacities comes into question.

In one sense lower-level adaptation occurs constantly, in the process Charles Lindblom calls "muddling through," by which existing policies are altered in a piecemeal way.[15] Keohane and Nye distinguish between "incremental" and "crisis-induced" learning, though the phenomenon they refer to is not learning as defined by Haas but adaptation in the sense I use here.[16] Muddling through equates with incremental adaptation. Crisis-induced adaptation involves more than just muddling through. Failures, breakdowns, and crises provoke leaders to search for new policies.[17]

In the search for new policy instruments decision-makers abandon the means but not the ultimate ends behind a policy that has been invalidated by experience. In this sense, as Martin Landau has shown, policies are like hypotheses and can be falsified.[18] In the case

13. This definition corresponds to that employed by Ernst Haas, who draws a similar distinction between learning and adaptation. The principal difference is that Haas applies the notions to international organizations and I apply them to national decision-makers. See Haas, *When Knowledge Is Power*, 33–34.

14. On hierarchies of ends and means in decision-making, see Simon, *Administrative Behavior*, chaps. 3–4.

15. Charles E. Lindblom, "The Science of 'Muddling Through.' "

16. Robert O. Keohane and Joseph S. Nye, Jr., "*Power and Interdependence Revisited*," 751.

17. By *crisis* I do not mean confrontations involving the threat of military force, as in the "crisis literature." Rather, I refer to a condition described by Ernst Haas as "the compounding of uncertainty in the minds of actors engaged in collective decision-making—uncertainty about the adequacy of cause-and-effect links carried over from past experience, about the proper ranking of values in competition, about the future toward which one should be working," though I emphasize the first element. See Ernst B. Haas, *The Obsolescence of Regional Integration Theory*, 25.

18. Martin Landau, "On the Concept of a Self-Correcting Organization."

of telematics collaboration in Europe, tinkering with the national-champion strategies involved incremental adaptation. Adding a collaborative component to policy displayed crisis-induced adaptation. Crisis-induced adaptation leads to purposive searching. I shall use *adaptation* as shorthand for crisis-induced adaptation from now on and shall call the searching it provokes *the adaptive mode*.

The difference between adaptation and learning consists of this: Learning has to do with the nature of the problem; adaptation has to do with the limits of the material and political resources of the state. The notion of adaptation implies that the failure of unilateral strategies provokes a rethinking of policy and a search for new approaches. When several states simultaneously face a problem of scale, potential demand for cooperation exists.

Leadership and International Organizations

The second part of this analytical schema confronts the question, Who organizes cooperation? Policy adaptation in a set of countries may create an opportunity for cooperation, as state leaders react to failed unilateral strategies by searching for new approaches. But cooperation does not necessarily emerge self-created out of the soup of failed unilateral strategies. Some political actor (or actors) must propose cooperation and sell the idea to potential collaborators. Such a political leader or entrepreneur must mobilize a coalition in favor of cooperation. Without a leader, the demand for cooperation is likely to remain latent.

Some theorists make the case that, in certain circumstances, cooperation can emerge spontaneously from the unorganized, self-interested behavior of rational actors. Axelrod demonstrates that cooperation can evolve, provided a sufficiently large core of actors chooses the right kind of strategy (tit-for-tat) and their interactions extend over an indefinite period of time.[19] This line of analysis has been elaborated and applied to a variety of international interactions. The contributors to the volume *Cooperation under Anarchy*, for example, emphasize three aspects of the structure of the strategic game as "favoring" cooperation: mutuality of interest (the structure of payoffs), iteration of the game, and the number of actors.[20]

19. Axelrod, *Evolution of Cooperation*.
20. Kenneth A. Oye, ed., *Cooperation under Anarchy*.

But none of the analysts developing or applying these notions claims that these game-theoretic dimensions are either necessary or sufficient for international cooperation to arise. And though it is certainly worthwhile specifying some conditions that make cooperation likely, their analyses leave open an immense array of other variables that are crucial in determining the effects of the game-theoretic factors. In fact, Axelrod and Keohane, in concluding *Cooperation under Anarchy,* point out that the context of an issue frequently has "a decisive impact on its politics and outcomes." They argue that the perceptions of actors and international institutions can be of particular import.[21]

The cooperation-under-anarchy approach needs shoring up in another way as well. In many (if not most) cases costs are associated with organizing cooperation. States must frequently be persuaded to appear at the discussion table. Someone must propose mechanisms for implementing cooperation and must broker compromises. Political leaders are actors who are willing to assume the costs of organizing. This insight derives from another (and earlier) branch of rational-choice theorizing, that which pioneered the analysis of public goods. As Norman Frohlich, Joe Oppenheimer, and Oran Young note:

> Except in the unusual case of the single individual who supplies a collective good, it is generally agreed that some sort of organization is required to collect resources and to supply the good in question. Yet discussions of collective goods seldom pay much attention to the process through which such an organization can or will come into existence in a social structure.[22]

Political leadership is the key to organizing cooperation.

Frohlich and his collaborators argue that organizing to provide collective goods entails certain costs. In other words, there are costs beyond those of providing the collective good itself, and those costs have to do with mobilizing participants. The cost of providing a collection organization can be assumed by a political leader or political entrepreneur. The political leader does not pay the costs of organizing out of altruism; she expects to derive some sort of surplus or net gain. The surplus could be psychic (like enhanced pres-

---

21. Robert Axelrod and Robert O. Keohane, "Achieving Cooperation under Anarchy," 227.
22. Norman Frohlich, Joe A. Oppenheimer, and Oran R. Young, *Political Leadership and Collective Goods,* 6.

tige) or pecuniary (material profits). The notions of political leadership and the costs of organizing have been developed with reference to domestic politics and policy processes but have not been exploited in the study of international politics.

To be sure, students of international politics have constantly analyzed international political leadership. But we have not always been cognizant of the costs of organizing cooperation and by whom those costs can be assumed. The predominant mode of thinking has been that single powerful states play the crucial role of organizing cooperation. In fact, this argument goes back as far as Mancur Olson, one of the pioneers in the analysis of collective action. Olson argued that public goods could be supplied when one powerful actor either valued the good more than the cost of providing it (and so supplied it for all) or could make private side payments in order to enlist contributions.[23] Kindleberger made the case that a liberal international economic order depended on the existence of a hegemon to lead it, such as Britain in the nineteenth century and the United States in the mid-twentieth.[24] Indeed, the notion of single-power leadership became the heart of the hegemonic-stability theory of international regimes.[25] Thus the economic preeminence of the United States was seen as explaining the creation and stability (into the 1970s) of the Bretton Woods system. More recently, scholars have shown that, in principle, a small number of states can jointly provide international leadership.[26]

The study of international cooperation requires a more generally useful notion of political leadership than that provided by hegemonic-stability theory. I propose revising the concept of international political leadership in two ways: first, by decoupling it from public goods so as to apply it to private goods as well; and, second, by including international organizations among the potential sources of political leadership. Before proceeding with those arguments, however, I will specify the functions of international political lead-

23. Mancur Olson, Jr., *The Logic of Collective Action*.
24. Kindleberger, *World in Depression*.
25. Key early statements were Robert Gilpin, *U.S. Power and the Multinational Corporation*; and Stephen D. Krasner, "State Power and the Structure of International Trade."
26. See, for example, Keohane, *After Hegemony*; Snidal, "Limits of Hegemonic Stability Theory"; David A. Lake, "Beneath the Commerce of Nations: A Theory of International Economic Structures."

ership in the initiation of cooperation. International political leaders perform some or all of the following functions:

1. *Proposing:* suggesting to potential collaborators that areas exist in which collective action would be mutually beneficial

2. *Mobilizing:* drawing potential collaborators into discussions over the proposed collaboration

3. *Shaping the agenda:* providing an initial framework for negotiation, including potential goals and modalities of cooperation

4. *Building consensus:* promoting common technical understandings of the problem and cause/effect relations

5. *Brokering compromises:* defining potential regions of agreement between divergent interests

All these functions require investments of time, personnel, energy, and financial resources. All relate to the earliest stage of initiating collaboration.[27]

One would derive a different notion of leadership from the traditional, realist-inspired literature on international politics. To be sure, scholarship on international politics does not generally employ explicit concepts of leadership. Rather, initiative in world politics is presumed to be the prerogative of the powerful. In effect, leadership means the ability to coerce or compel other actors to go along with what the "leader" desires. Leadership is implicitly linked to power, whether power means control of resources that others need or the ability to obtain desired outcomes in the face of opposition.[28] Keohane and Nye presumably had in mind this notion of leadership when they wrote that "leadership will not come from international organizations, nor will effective power."[29]

There are, however, other notions of leadership. Leadership need not imply the use of power to coerce or compel. At one extreme, in fact, leadership can rest wholly on the charisma and personal appeal of an individual—entirely without coercion or side pay-

27. The initiation of collaboration is the focus of this study. Other important functions and variables may be involved in maintaining or managing cooperation once it is underway, but they require separate development.
28. See David Baldwin, "Interdependence and Power: A Conceptual Analysis."
29. Robert O. Keohane and Joseph S. Nye, Jr., *Power and Interdependence,* 240.

ments.[30] Students of organizations have long recognized different kinds of authority and leadership. For instance, Amitai Etzioni proposed that within organizations there could be different "means of control": coercive, utilitarian, and social.[31] Etzioni's definitions may not be directly applicable to relations among states, but they certainly suggest that organizational interactions, even at the international level, might involve forms of leadership other than those based on power and coercion.

Keohane and Nye, though they reject international organizations as sources of power or leadership, suggest that they can exercise initiative. They argue that "international organizations may therefore help to activate 'potential coalitions' in world politics, by facilitating communication between certain elites; secretariats of organizations may speed up this process through their own coalition-building activities."[32] I suggest that this kind of activity on the part of international organizations constitutes international leadership. It is leadership defined by the functions outlined above: persuading, setting agendas, mobilizing coalitions, promoting consensus, and pushing compromises. Such noncoercive leadership can, I will argue, be effective in initiating international cooperation.

As we have seen, the theoretical literature dealing with leadership and collective action grew around the problem of public goods. Pure public goods satisfy two criteria: jointness of consumption, meaning that consumption of a unit of the good by one actor does not reduce its availability to others; and impossibility of exclusion, meaning that once the good is provided, consumption cannot be limited to those who contributed to its provision. The public-goods theorists showed that these characteristics mean that public goods will always be underprovided, if they are provided at all. In this theory, political leadership is the means by which public goods can be supplied and the costs can be shared.

The analytical shortcoming with this approach is that pure public goods are rare in the real world. Indeed, many instances of international cooperation probably involve the provision of goods that cannot be jointly consumed. Furthermore, most international collective action deals with goods for which exclusion is feasible and

---

30. See Max Weber, *From Max Weber: Essays in Sociology.*
31. Amitai Etzioni, *Modern Organizations*, chap. 6.
32. Keohane and Nye, *Power and Interdependence*, 240.

routinely practiced. The free-trade regime centered on the General Agreement on Tariffs and Trade (GATT)—presumably one of the most nearly "public" of international goods—excludes scores of nations.

I propose that costs of organizing exist even for private goods that require cooperation for their provision. In other words, even for goods that are not joint and for which exclusion is feasible, there is still a problem of collective action whenever a single actor cannot provide the good for herself. This point has also been made by Russell Hardin, but it is frequently overlooked.[33] The upshot is that the analytical framework I propose in this study, centered on cognitive adaptation and political leadership, should be useful for any kind of international collective action, for public or private goods. A latent demand for cooperation will not become collective action until a political leader assumes the cost of organizing cooperation. And the costs of organizing exist for private as well as public goods: Someone must propose, persuade, mobilize, prepare an agenda, and broker compromises.

For the cases of European technological collaboration, this point becomes important. In principle, technological knowledge is joint— my possession of it does not diminish the amount of it available for you. However, in practice, the more possessors there are of technological knowledge, the more competitors there are likely to be in the relevant commercial market. Market shares are not joint or are at the least extremely prone to crowding. Thus technological knowledge is one short step away from being nonjoint. Companies tend to think of technical information as nonjoint and hence seek to protect it. Patents and copyrights exist for precisely that purpose. Thus the broad notion of political leadership I propose is crucial for the empirical cases analyzed here and for all other instances of collective action for nonpublic goods.

The analytical reach of the notion of international political leadership can be expanded in another way. The potential sources of leadership in the international arena are broader than is usually recognized. International leadership can come from single powerful states or small groups of states. But international organizations (IOs) can, under certain circumstances, also exercise the leadership func-

33. Russell Hardin, *Collective Action*.

tions outlined above. This flatly contradicts the realist assertion that "international institutions are unable to mitigate anarchy's constraining effects on inter-state cooperation."[34]

My argument also contradicts the neorealist hypothesis advanced by Andrew Moravscik, who argues that EC institutions had no impact on the definition of interests or the political bargaining that produced the Single European Act (SEA). Moravscik contends that the SEA was the product solely of bargaining among France, Germany, and the United Kingdom. Unfortunately, this hypothesis is at once obvious and misleading. Ultimately, all important decisions in the Community involve bargains among EC states, and the large states have a greater say in those bargains than do the small states. But Community institutions can shape the agenda, promote the redefinition of interests, and take the lead in proposing solutions to common problems. In the case of the SEA, the European Parliament started the process. The Parliament's Draft Treaty placed European union on the agenda for national governments; national parliaments were examining and voting on it. Thus Mitterrand's campaign for renewal of the Community did not emerge out of a void; the French president attached his prestige and his ambitions to the Parliament's proposals. Moravscik states that the Commission's White Paper "was a response to a mandate from the member states"; but, in fact, the member states were only responding to the initiative of the European Parliament.[35]

One important line of theorizing is that international institutions in general (including regimes as well as specific IOs) can sustain cooperation by reducing the costs of transactions and information. Reduction of transaction costs lays a basis for reciprocity and stable expectations about behaviors.[36] Oran Young argues that IOs can be a source of innovative policy ideas and can act at times almost like pressure groups.[37] My proposition is similar but goes still farther. Under some conditions IOs can exercise political leadership and influence the adaptive behaviors of states. Reducing transaction and information costs are relatively passive functions; IOs can also actively lead.

34. Grieco, "Anarchy and the Limits of Cooperation," 485.
35. Andrew Moravscik, "Negotiating the Single Act: National Interests and Conventional Statecraft in the European Community."
36. A clear statement of this argument is in Keohane, *After Hegemony.*
37. Oran Young, *International Cooperation: Building Regimes for Natural Resources and the Environment,* 54.

What are the conditions under which international leadership can come from IOs? First, the greater the initial grant of decision-making authority to the IO, the greater its ability to lead in new areas. IOs with limited mandates and specific issues will not have much latitude for leadership; the International Civil Aviation Organization is an example of this type. The Commission of the European Communities is one of the IOs best endowed with some autonomous capacities (for example, competition policy) and broad powers to propose.

Second, when representatives of an IO have substantive knowledge and relevant information, they can help shape the technical discussions and agreements. One of the most dramatic examples of this effect is the economic analysis generated by Raul Prebisch and his fellow economists at the United Nations Economic Commission for Latin America (ECLA). ECLA thinking was incorporated into the philosophical foundations of Latin American regional organizations, as well as the United Nations Conference on Trade and Development. In the case of telematics in Europe, the Commission was armed with technical studies and data that allowed Commission representatives to structure and contribute to the technical discussions leading to ESPRIT and RACE. The Commission acted as an informed participant in the discussions, influencing the direction of discussions and the consensus that emerged.

Third, IOs can exercise initiative, proposing solutions to common problems and rallying support among member governments. Their ability to perform these functions depends in part on the personal characteristics of IO leaders. As Robert Cox and Harold Jacobson suggest, IO leaders possess different sources of influence, some inhering in their position (legitimacy, authority) and some deriving from personal characteristics. Personal attributes that enhance the influence of an IO official include charisma, expert knowledge, negotiating ability, personal achievement outside the IO, and administrative competence. Thus, "high international officials may command information and recognition, which allows them the initiative in proposing action or resolving conflict."[38]

Fourth, and most important, I submit that international organizations will register the greatest impact on interstate cooperation during periods of crisis (as defined previously in this chapter). When

38. Robert W. Cox and Harold K. Jacobson, *The Anatomy of Influence,* 20.

national leaders confront policy failures that compel them to re-think their objectives or the means chosen to pursue them (or both), they are in the adaptive mode—searching for alternative ap-proaches. IO actors can seize the initiative under such circum-stances to supply new models and strategies. In addition, crises pro-vide opportunities for activist, entrepreneurial IO leaders to marshal states behind a cooperative solution. The activist IO leader not only can sponsor transnational contacts and consensus building but can sometimes appeal directly to heads of state. In other words, the efforts of IO officials will count for more when member countries are simultaneously in an adaptive mode. Ernst Haas provides ex-amples of this effect in the United Nations.[39] In this study Com-missioner Davignon plays the role of the entrepreneurial IO leader.

Under these four conditions—broad initial grant of authority, technical preparedness, activist IO officials, and policy-making cri-sis in member countries—IOs enjoy the broadest possible oppor-tunity to exercise international political leadership. Political lead-ership, as I have argued, depends on the leader's receiving some form of net benefit. For IOs the gains take the form of enlarged mandates, increased prestige, perhaps even expanded budgets and staffing. With regard to paying the costs of organizing new coop-erations, IOs possess certain advantages over states. What may be a cost to states trying to organize cooperation—in time, resources, energy—is a matter of course to IOs. IOs exist to foster coopera-tion. In other words, what may be a cost to a national government is a part of the mission of IOs.

The Problem of *Juste Retour*

The third element in this theoretical framework bears particularly on instances of collective action for private goods. I take it as given that states will seek assurances of a fair return on their investment in the cooperation. The general theoretical point, enunciated by James March and Herbert Simon with respect to all organizations, is that each member expects a balance between contributions and rewards, between what it puts into and what it receives from the organiza-tion.[40] I propose that governments in organized cooperation expect

39. Haas, *When Knowledge Is Power*.
40. James G. March and Herbert Simon, *Organizations*. I am grateful to Judith Haymore Sandholtz for bringing this line of theorizing to my attention.

the same. Governments are unlikely to agree to collaborate unless they have some confidence in obtaining a satisfactory ratio of contributions to benefits. What constitutes a satisfactory ratio is entirely subjective. Furthermore, the *juste retour* problem is not one with a unique solution.

To begin abstractly, two extremes bound a continuum of solutions. Lindblom suggests that problems of resource allocation can have market and political solutions.[41] Though Lindblom writes of national economies, a similar logic is relevant here. At the market extreme, individual choices interact to produce a distribution of resources; at the opposite extreme, a central authority decrees the distribution.[42] In technological collaboration, the market approach would be the pure à la carte system, in which countries bid to participate only in those parts of a program in which they are interested. At the political end of the continuum, a central authority would assign agreed-upon shares.

In practice, both forms of allocation exist. Pure à la carte participation is the basis of the EC's Coopération Scientifique et Technologique (COST) program and of the EUREKA program: States participate only in projects of their choosing. Indeed, the programs are really umbrellas for intergovernmental R&D agreements. In contrast, shares of work in the Airbus consortium are set by contract. Similarly, the members of the European Space Agency (ESA) agreed in January 1985 to a hybrid approach: Each country should receive industrial contracts worth 95 percent of its financial contribution, yet countries participate in only those major programs that they choose to.[43]

I hypothesize a straightforward connection between the method of handling *juste retour* and the mission of the cooperative effort. Collaboration for a single, discrete aim will allocate shares authoritatively because later adjustments are likely to be difficult. Collaboration for a large and differentiated set of objectives will be likely to distribute shares by self-selection. If a collaborative effort is ongoing (iterated) and composed of myriad bits and pieces, states can pick and choose which ones they want to participate in and still be confident, over the long run, of achieving a fair share. A one-shot project cannot offer that assurance; the initial shares will also be

41. See Lindblom, *Politics and Markets*.
42. See ibid.
43. I examine ESA at length in Chapter 5.

the final shares, barring a painful renegotiation. Put differently, the more pies, the easier it is to see that everyone gets enough slices of the right sizes.

In technological collaboration single-purpose projects will be likely to display authoritative allocation of shares. This is certainly the case with Airbus and with cross-national jet-engine collaboration. Each participant receives a set share, negotiated at the outset. In contrast, programs that aim to advance technological capacities across a broad front (rather than to develop a specific item) will display à la carte participation. This is the case in programs like ESPRIT and EUREKA.

SUMMARY

Taken together, these three elements—the demand for cooperation, the supply of political leadership, and the allocation of fair returns—provide a theoretical framework for analyzing international cooperation. The framework is a strategy for analysis. Because, by assumption, states prefer autonomy to cooperation, analysis must account for the cognitive process by which decision-makers decide that cooperation is palatable. I have proposed two broad kinds of international problems, each linked to a different species of cognitive change. Solutions to problems of scope inherently require coordinated action on the part of multiple states. Cross-border pollution is a problem of scope. For such problems, learning is the relevant cognitive process. Policy preferences change as decision-makers acquire increased technical understanding of the problem.

With problems of scale, the key question deals with resources. Does a state have the material and political resources to pursue unilateral solutions? Two gaping uncertainties work against a priori answers to that question: There is frequently no objective threshold of material resources, knowable in advance; and so much depends on the ability of the state to appropriate and channel resources. The upshot is that states must generally discover through experience the boundaries of unilateral action. Especially when the objective is highly valued, states will attempt unilateral approaches first and adjust policies in response to experience. This is the process of adaptation.

In either case, once decision-makers conclude that unilateral strategies are inadequate, the demand for cooperation is still only potential. Indeed, I have proposed that there are costs of organizing

TABLE 2.1.   ANALYTICAL FRAMEWORK FOR
EXPLAINING INTERNATIONAL COOPERATION

I. **Cognitive variables:** the demand for cooperation
   A. Problems of scope → learning
   B. Problems of scale → adaptation
      1. Failure of unilateral strategies
      2. States in adaptive mode
II. **Costs of organizing:** the supply of leadership
   A. Hegemonic leadership
   B. Small groups of states in joint leadership
   C. International organizations
III. **Arrangements for *juste retour***
   A. Single-purpose cooperation → shares by binding agreement
   B. Multipurpose cooperation → à la carte participation

cooperation that some actor or actors must assume. Assuming these costs is the function of international political leadership. I have further proposed that under certain circumstances international organizations can exercise political leadership. They can propose, persuade, mobilize, set the agenda, and broker compromises. This proposition runs counter to much realist and postrealist thinking, which attributes to IOs either passive functions (reducing transaction and information costs) or no important functions at all.

The third element in this theoretical framework refers directly to private goods and concerns arrangements to address the problem of fair returns. I argue that states are not always and everywhere preoccupied with preventing other states from achieving relative gains. I reject the (unverifiable) assertion that states translate all international interactions into a calculus of gains and losses of power relative to other states. Rather, I propose that states demand a balance between their expected gains from cooperation and their contributions to the effort.

Taken together, the cognitive, leadership, and fair-return variables provide a general framework for analyzing international cooperation. Table 2.1 summarizes the three main elements of the theoretical framework and some of their fundamental implications. The test of any theoretical construct is its ability to support satisfying explanations of a range of empirical phenomena. In the chapters that follow, I analyze in detail three major European collaborative technology programs of the 1980s by using these notions. Furthermore, I provide condensed summaries of other joint European technology ventures. The cases cover all the major instances of postwar European technological collaboration in the civilian sector. The features common to all the cases (European, civilian high technology, postwar) limit the number of variables so as to make comparisons meaningful. By the same token the cases differ in technologies addressed, organizational settings, and historical context.

My overarching analytical aim, then, is to illuminate the politics of international cooperation. European high-technology programs are, in that context, the raw materials for analysis. Fortunately, they also make for fascinating stories in themselves. The dramatic elements of international politics are all present: the quests for power and wealth; conflict, competition, and compromise; failure and, sometimes, success.

THREE

The Telematics Revolution

The word *revolution* has probably been emptied of meaning by its too frequent use; even a new laundry detergent can be labeled revolutionary. Still, occasionally we observe a revolution that really is "a complete, pervasive, usually radical change in something."[1] In fact, the world is in the midst of a period of technological change that fits that definition. This revolution grew from startlingly new ways to control and employ electricity: integrated circuits (ICs). A single silicon chip can now perform the functions that in the mid-1950s would have required a computer filling a large room. That chip (the IC) is the basic building block of larger systems that process, store, and communicate information.

Virtually no aspect of life today is sheltered from the telematics revolution. Telematics has transformed manufacturing, transportation, communications, banking and finance, entertainment, retailing, and defense. In the industrialized world, though every country feels the changes spawned by telematics, some countries benefit more than others from producing and diffusing the technical innovations that keep the revolution rolling. Naturally, technology has been politicized. Politicians, prime ministers, presidents, and parliaments all fret over the economic consequences for their country of this latest technological revolution.

---

1. *The Random House College Dictionary,* 1973, s.v. "revolution."

In this chapter I introduce the telematics technologies and show that they are at the core of a technological revolution. I begin by laying out basic definitions for terms that will appear throughout the study, like *technology, high technology,* and *innovation.* The ensuing argument begins broadly with a discussion of technology and economic change. I describe a world in which technological change is constant but in which clusters of radical innovations occasionally trigger massive changes in the economy. The case for a telematics revolution is built on the rapid growth of the sectors themselves (semiconductors, computers, and telecommunications) and on their far-reaching repercussions throughout the economy.

CONCEPTS AND DEFINITIONS

Everyone probably has an intuitive notion of what science and technology are. Unfortunately, sometimes intuitive ideas are not very helpful. For instance, it might seem logical to suppose that technology is the application of scientific knowledge to solve real-world problems. But this view obscures more than it clarifies. Technology comprises knowledge of its own, quite independent of scientific understanding. I take science to be the systematic quest for knowledge, especially general laws and principles, about the universe. Science is systematic because its practitioners share beliefs about what needs to be explained and about valid methods for creating hypotheses and evaluating evidence.

Technology is a different kind of knowledge, not necessarily derived from the discoveries and laws of science. Technology embodies knowledge about "techniques, methods, and designs that work, and that work in certain ways and with certain consequences, even when one cannot explain exactly why."[2] Or, put differently, technology in general means "the knowledge to design goods and services to attain certain objectives." A specific technology, then, comprises a "set of design principles, based on technological or scientific understanding, which identify particular commodities and their processes of production."[3] This definition of technology highlights the complex relation between science and technology. Nathan Ro-

2. Nathan Rosenberg, ed., *Inside the Black Box: Technology and Economics,* 143.
3. J. S. Metcalfe, "Technological Innovation and the Competitive Process," 38.

senberg captures rather neatly the major dimensions of that relation, and I summarize his treatment in the next paragraphs.

First, technology frequently outruns science. Because economic
benefits sometimes reward technological innovation, technical en-
hancement derived from technological knowledge often precedes
scientific understanding of the processes involved. For instance, engineers at Bell Labs were using semiconductors like copper oxide
and silicon rectifiers far in advance of the scientific (solid-state physics)
research that illuminated the various physical processes underlying
them.

Second, even in industries built on scientific knowledge—like
electronics—improvements based on user demand and engineering
experience are common. These enhancements sometimes provide
the raw material or stimulus for scientific investigation. Thus, scientific advances sometimes occur in the process of trying to improve
the performance of existing products. Rosenberg cites the discovery
of star noise by researchers trying to identify sources of noise in
transoceanic radiotelephone transmission.

Finally, scientific discoveries owe much to equipment and tools
developed for practical uses. Instruments developed to meet specific
engineering problems enable scientists to observe and measure new
phenomena. The most striking example is certainly the computer,
which allows scientists today to make immense calculations that a
generation ago would have been impossible. But microscopes, tele-
scopes, and chromatographs also fit the pattern. In short, technology feeds science, and science feeds technology. Sometimes scientific discoveries lead to new technologies and products; sometimes
new technologies stimulate and enable progress in science. Science
and technology are two highly interactive kinds of knowledge.[4]

In these days of constant and rapid technological change, we have
learned to distinguish between high technology and other kinds of
technology. A number of criteria could serve to distinguish high-
technology sectors: R&D intensity, rapid obsolescence of products
and processes, high risk and uncertainty, strategic priority for governments, high level of patenting activity. The first of these, R&D
intensity, is the easiest to use, largely because the Organization for
Economic Cooperation and Development (OECD) has gathered data
on R&D expenditures and personnel by business and governments,

4. Rosenberg, "How Exogenous is Science?"

broken down by broad categories. Even better, the OECD data are comparable across countries because the procedures by which member countries report are standardized and formalized in the periodically revised "Frascati Manual."[5]

On the basis of these data the OECD has calculated the ratio of R&D expenditure to output for a number of industries; the resulting figures are the weighted average for a set of eleven OECD countries (taken as a single area). Table 3.1 shows the low, medium, and high R&D-intensity industries in 1970 and 1980. Interestingly, the same six industries constituted the high category in 1980 and in 1970. Furthermore, they increased their average level of R&D intensity, largely because of those sectors that involve telematics (office machines and computers, electronics and components).

This classification method may not convey the importance of certain high technologies for which R&D is spread over a number of categories. I am thinking in particular of new materials (being pursued in the aerospace, automobile, chemicals, plastics, glass, and metals industries) and biotechnologies (emerging in the drug and agricultural industries but also in small genetic-engineering start-ups). Still, the telematics industries are clearly among the highest of the high tech, and their research intensity is the fastest growing.

We can also draw distinctions among high-tech sectors according to their economic importance. Some technologies, like microelectronics, are vital because they make possible enhanced productivity and qualitative improvements (for example, better performance) in a broad range of industries. Following Richard Nelson, I refer to technologies that fit this description as leading technologies. Nelson speaks of industries whereas I speak of technologies, but the idea is the same. Nelson's leading industries are those that "drive and mold economic progress across a broad front." In contrast to leading technologies, strategic technologies are those that governments decide are worthy of public support because "national economic progress and competitiveness are dependent upon national strength in these industries."[6]

5. The official document is OECD, *The Measurement of Scientific and Technical Activities: Proposed Standard Practice for Surveys of Research and Development.*
6. This is the formulation of Richard R. Nelson, *High-Technology Policies: A Five-Nation Comparison*, 1. In practice, strategic technologies are almost always leading technologies, though they need not be.

A final set of concepts to be defined centers on technological change. First, it should be clear that technological change occurs all the time, in ways as "small" as modifying the jig for a machine tool or as "big" as discovering a way to make synthetic fibers. In other words, technological change is both incremental and disjunctive. It occurs in laboratories, design offices, garages, and on shop floors. Not only researchers but also manufacturers and users of products can be responsible for technological changes. Second, there is a difference between invention and innovation. Invention is the original conception of a new product or process or of a technical improvement in an existing product or process.[7] Innovation is "the technical, financial, manufacturing, management and marketing activities involved in the commercial introduction of a new product, or in the first commercial use of a new manufacturing process or equipment."[8] Innovation is thus a much broader concept than invention, as it involves the development, production, and commercialization of an invention. Many inventions never get farther than the idea stage; innovations enter the economy and begin to diffuse among producers and consumers.

We can also distinguish among degrees of technological change. Christopher Freeman proposes a useful typology:

1. *Incremental innovations:* continuous changes, resulting from R&D at times but also from production experience and user initiative, and leading to improvements or efficiencies in existing products and processes

2. *Radical innovations:* often the result of R&D coupled with entrepreneurial activity; often occur in clusters; cannot happen incrementally because they replace some goods or processes. An example is the emergence of nylon.

3. *Technology-system change:* combined effect of many radical and incremental innovations and organizational changes affecting

7. A *product* is a good or service that has been created by transforming certain inputs (raw materials, energy, know-how) into output. A *process* is the set of methods and techniques used to transform the inputs into outputs. An IC is a product; the planar process (for making silicon wafers with an insulating oxide layer), lithography, etching, and doping are all processes involved in making ICs. Whether a particular development involves product or process depends on one's point of view. X-ray lithography equipment involves a product change for makers of semiconductor-manufacturing equipment, but a process change for makers of semiconductors. This point is summarized nicely in Rosenberg, *Inside the Black Box,* 4.

8. Roy Rothwell, "Reindustrialization, Innovation and Public Policy," 67–68.

TABLE 3.1. R&D INTENSITY IN THE OECD AREA,
WEIGHTING OF THE ELEVEN MAIN COUNTRIES,
R&D EXPENDITURE/OUTPUT

	1970		1980	
		Intensities		*Intensities*
High			**High**	
1. Aerospace		25.6	1. Aerospace	22.7
2. Office machines and computers		13.4	2. Office machines and computers	17.5
3. Electronics and components		8.4	3. Electronics and components	10.4
4. Drugs		6.4	4. Drugs	8.7
5. Instruments		4.5	5. Instruments	4.8
6. Electrical machinery		4.5	6. Electrical machinery	4.4
	Average	10.4		Average 11.4
Medium			**Medium**	
7. Chemicals		3.0	7. Automobiles	2.7
8. Automobiles		2.5	8. Chemicals	2.3
9. Other manufacturing industry		1.6	9. Other manufacturing industry	1.8

10.	Petroleum refining	1.2
11.	Nonelectrical machinery	1.1
12.	Rubber and plastics	1.1
	Average	1.7

Low

13.	Nonferrous metals	0.8
14.	Stone, clay, glass	0.7
15.	Shipbuilding	0.7
16.	Ferrous metals	0.5
17.	Fabricated metal products	0.3
18.	Wood, cork, furniture	0.2
19.	Food, beverages, tobacco	0.2
20.	Textiles, footwear, leather	0.2
21.	Paper, printing	0.1
	Average	0.4

10.	Nonelectrical machinery	1.6
11.	Rubber and plastics	1.2
12.	Nonferrous metals	1.0
	Average	1.7

Low

13.	Stone, clay, glass	0.9
14.	Food, beverages, tobacco	0.8
15.	Shipbuilding	0.6
16.	Petroleum refining	0.6
17.	Ferrous metals	0.6
18.	Fabricated metal products	0.4
19.	Paper, printing	0.3
20.	Wood, cork, furniture	0.3
21.	Textiles, footwear, leather	0.2
	Average	0.5

SOURCE: Organization for Economic Cooperation and Development, *R&D, Invention and Competitiveness*, OECD Science and Technology Indicators, No. 2 (Paris, 1986), 59.

one or more sectors of the economy and creating new sectors; for example, the interconnected innovations involving synthetic materials, petrochemicals, injection molding, and extrusion that were developed from the 1930s through the 1950s

4. *Technoeconomic-paradigm change:* changes in technology systems that are so pervasive they affect the entire economy. These constitute "technological revolutions" that combine "radical product, process and organisational innovations" and occur only once or twice a century. The changes are so far-reaching that they entail the adjustment of industrial, social, and political systems. Examples include steam power and, most recently, electronics.[9]

At this point we have an intersection between the previous definition of leading technologies and this discussion of degrees of technical change. Clusters of leading technologies are the driving force behind Freeman's technological revolutions.

Finally, we must touch briefly on the question of the motive force behind technological innovation. Two broadly styled kinds of forces have been advanced as the prime motors. The first is the technology push, or the supply side. The argument runs that discoveries and inventions by researchers and technologists present new opportunities, which are then developed and diffused as markets are opened for them. The second is the market pull, or the demand side. Here, the needs identified by users and consumers stimulate the search for technological answers; thus, innovation is a response to the market. In fact, it seems virtually certain that both factors matter. As David Mowery and Nathan Rosenberg conclude, "Rather than viewing either the existence of a market demand or the existence of a technological opportunity as each representing a sufficient condition for innovation to occur, one should consider them each as necessary, but not sufficient, for innovation to result; both must exist simultaneously."[10]

I have no intention of recapitulating the arguments and evidence surrounding this issue. Rather, I accept as sufficient for my purposes the conclusion that both technology supply and market demand matter for innovation; I leave to economists the specification

9. Christopher Freeman, *Technology Policy and Economic Performance: Lessons from Japan,* 61–76.
10. David C. Mowery and Nathan Rosenberg, "The Influence of Market Demand upon Innovation: A Critical Review of Some Recent Empirical Studies," 231.

of how much and under what circumstances. But the distinction will be relevant when I discuss government policies. It will emerge (in the next chapter) that European leaders in the 1950s and early 1960s adopted science policies based on a (probably implicit) technology-push view. Beginning in the late 1960s they began to recognize the importance of markets, and technology policies began to reflect a concern for market-pull factors, like diffusion and applications.

TECHNOLOGICAL CHANGE AND ECONOMIC GROWTH

Real-world events of the last few decades have lent new relevance to Joseph Schumpeter's insight that technological change is the real heart of economic competition. Whereas classical economic theory holds that firms compete on the basis of price, Schumpeter recognized that the real contest between harness makers and car manufacturers had less to do with relative prices than with technological changes that shifted the entire production function. In this section I spin out some of the implications of Schumpeter's thinking and cite the experience of Japan as confirmation of those implications.

As explained by Rosenberg, most economists discuss technical change "as if it were solely cost-reducing in nature, that is, as if one could exhaust everything of significance about technical change in terms of the increases in output per unit of input that flow from it." What this view misses, argues Rosenberg, is that economic growth in the West has owed more to qualitative changes in the kinds of goods and services produced than to reducing the cost of existing products. Put another way:

> Western industrial societies today enjoy a higher level of material welfare not merely because they consume larger per capita amounts of the goods available, say, at the end of the Napoleonic wars. Rather, they have available entirely new forms of rapid transportation, instant communication, powerful energy sources, life-saving and pain-reducing medications, and a bewildering array of entirely new goods that were undreamed of 150 or 200 years ago.[11]

Thus, one of Schumpeter's main contributions to current debates is the proposition that technological change can alter not just production costs but the very terms of economic competition and growth.

11. Rosenberg, *Inside the Black Box*, 4.

Schumpeter went one step further to propose that groups of interlinked technological innovations trigger major disjunctions in the speed and direction of capitalist economic growth. Such technological revolutions, in his view, are behind the long business cycles of growth and decline identified by Nikolai Kondratieff. Freeman enumerates five technological revolutions associated with Kondratieff cycles:

1. *1770–1840:* early mechanization, involving water power and especially affecting textiles, iron, and pottery

2. *1830–1890:* steam power, iron, and railways; also, machine tools and steamships

3. *1880–1940:* electrical and heavy engineering; electrical machinery, steel ships, heavy chemicals

4. *1930–1990:* Fordist mass production; autos and aircraft, consumer durables, synthetic materials, petrochemicals

5. *1980–:* information and communication technologies based on microelectronic chips; computers, electronic capital goods, telecoms, robotics and flexible manufacturing, information services[12]

The implication for international relations is portentous: During technological revolutions, the relative economic strength of nations can shift, as states better able to exploit the new ways of producing wealth surpass the old leaders.[13]

Two implications of Schumpeter's fundamental ideas can now be pursued. These implications contradict certain notions from classical economics: one about the determinants of comparative advantage, and one about the efficacy of government interventions. First, at the level of nations, factor endowments (a nation's supply of raw materials, capital, and labor) do not alone determine which products states can most benefit from producing and trading. Classical trade theories hold that a nation will do best by selling on world markets those products for which it has a comparative advantage—that is, those products that require input ratios corresponding to the nation's factor endowment. Thus nations with rel-

---

12. Freeman, *Technology Policy,* 66–71.
13. Some scholars have fashioned theories of international political-economic change around this insight. One notable example is Robert Gilpin, *The Political Economy of International Relations.*

atively more capital than labor will "naturally" specialize in capital-intensive manufactures. In traditional trade theories technologies simply enhance the productivity of other factors.

In contrast, recent theoretical explorations show that technology can create comparative advantage. Goods in high-tech industries are not differentiated solely by price; they differ in performance and capability. Countries that are first to exploit technological innovations can capture large profits during the phase in which no competition exists. As long as the technological advantage holds, there are increasing returns to scale, contradicting the basic assumption of traditional trade theory. Furthermore, continuous innovation through learning-by-doing enables countries to make qualitative jumps to new technologies.

Traditional economic theories would have the know-how and innovations diffuse across national boundaries at low cost, making technological advantages short-lived. But recent work shows that much technological knowledge is embodied in local networks of people and organizations and is therefore untradeable. Under such circumstances technological advantages can persist. In addition, certain leading sectors produce economic spillovers, transforming products and processes in a variety of industries. Semiconductors are a perfect example. Thus, advantages in leading sectors can produce competitive advantages in a host of economic activities.[14]

A second result of contemplating technological change and competition as Schumpeter suggested is that to the extent technological change can be encouraged or channeled by government policies, states can improve their competitive standing. This proposition is spun out by a number of students of technology policies, competitiveness, and the new trade theory who acknowledge Schumpeter as their intellectual ancestor.[15] They argue that productive advantage is malleable, and state policies can therefore "create" advantage.[16] The classical economic theories of trade hold that given de-

14. See Laura D'Andrea Tyson, *Creating Advantage: Strategic Policy for National Competitiveness;* and Giovanni Dosi, Laura D'Andrea Tyson, and John Zysman, "Trade, Technologies, and Development: A Framework for Discussing Japan."

15. See, for example, P. H. Hall, "The Theory and Practice of Innovation Policy: An Overview," 7–8; Freeman, *Technology Policy,* 1; and Stephen S. Cohen and John Zysman, *Manufacturing Matters: The Myth of the Post-industrial Economy,* 90.

16. The notion of creating advantage informs much of the work carried out by researchers at the Berkeley Roundtable on the International Economy. See, for ex-

clining returns to scale and perfect competition, all nations will benefit from free trade, and trade flows will automatically reach balance. Government intervention can only distort market signals and produce inefficiencies and net welfare losses. In fact, worrying about national competitiveness is pointless because every state will achieve trade balance based on exports of those goods for whose production it has a comparative advantage.

Japan refuses to conform to this theory. The Japanese government violated the maxims of classical economics by intervening to favor specific industries and sometimes specific firms, to restrict foreign investment, and to regulate the acquisition of foreign technologies.[17] In this fashion Japan has succeeded in graduating from technological imitation (albeit superb and always innovative) to technological leadership in many of the most advanced areas. As Freeman points out, Japan's Ministry of International Trade and Industry (MITI) chose for encouragement "the most advanced technologies with the widest world market potential in the long term."[18] Or, as Laura Tyson and John Zysman put it, Japan reached the technological forefront by virtue of an "interventionist targeting strategy toward those sectors having the greatest growth and technological potential."[19]

Thus, European decision-makers debate telematics collaboration in a Schumpeterian world in which technological innovation is an important key to economic growth and competition. In this world states can alter the competitive advantages of their industries. Indeed, one of the distressing problems facing Europe is the success of Japan in creating advantage by exploiting the opportunities created by the telematics revolution. The question for the European countries is whether competitiveness can be enhanced at a regional level. The uniqueness of the European collaborative effort in telematics lies in its attempt to achieve regionally what has been accomplished up to now only within nations: using high technology to build economic competitiveness.

ample, Laura D'Andrea Tyson and John Zysman, "American Industry in International Competition," 24–28; and, more recently, Cohen and Zysman, *Manufacturing Matters,* pt. 3, "Creating Advantage."

17. See Dosi, Tyson, and Zysman, "Trade, Technologies, and Development."
18. Freeman, *Technology Policy,* 35.
19. Laura D'Andrea Tyson and John Zysman, "Preface: The Argument Outlined," xvii.

THE TELEMATICS REVOLUTION

The word *telematics* is the English adaptation of a French neologism, a reversal of the trend sometimes decried by the French for English words to creep into and corrupt French usage. *Télématique* was coined to capture the convergence of two technologies: telecommunications and *informatique,* or data-processing. These two sectors, formerly as separate as telephones and adding machines, have now developed a significant area of overlap because of the miraculous operations of the IC. The IC makes possible the complex, rapid, electronic functioning of telecommunications and computing systems, and their increasing interweaving.

The Technologies

The marvel of ICs is that they contain, on a single chip of silicon approximately one centimeter square, hundreds of thousands of miniaturized electronic components (such as transistors and capacitors).[20] Two American firms, Texas Instruments and Fairchild, invented the IC in the late 1950s by implanting two or more miniature transistors on a single silicon substrate.[21] Today there are myriad types of ICs, though for my purposes here they can be classified into two major groups: logic circuits and memory circuits. Logic circuits are those that perform the logical operations within a system; they run the programs and manage the flow of electronic information. Logic chips that execute programs are called *microprocessors.* They function in conjunction with memory chips and input/output devices (which connect the system to its human user). Memory chips store electronic information, be it programs, operating instructions, or data. Some ICs contain both a processing unit and memory, and are called *microcomputers.*

Since the advent of ICs, chipmakers have refined their skills and steadily reduced the width of the metal paths connecting devices;

20. Silicon is the preferred semiconducting material for ICs, although for some uses (requiring the ability to withstand rough treatment or environmental extremes, for example) gallium arsenide serves better.

21. The silicon substrate is a wafer of silicon with an insulating oxide surface. The wafer is typically less than a millimeter thick. Devices (transistors, capacitors) are built into the wafer by implanting impurities (like boron and phosphorous) in selected places in the silicon. Patterns of circuitry, connecting the various devices, are created by etching patterns in the oxide layer and in layers of metal that are attached to it.

today this line width is less than one micron, or about one-tenth the diameter of a red blood cell. Consequently, the number of devices that can be crammed onto a single chip (the level of integration) has grown geometrically: today it is in the millions. The market for one-megabit (over 1,000K) ICs is presently growing, and four-megabit chips are commercially available. In the world of microprocessors the size of the word (or byte) that the circuit can move and manipulate has increased. The first microprocessor, introduced in 1971, could work with four-bit words.[22] It was quickly followed by the eight-bit processor, then the sixteen-bit processor (in 1973), and the thirty-two-bit (in 1980). Each generation of microprocessor is faster and more powerful than the previous one, making final systems correspondingly more capable.[23]

On top of the speed and capacity gains ICs achieved greater reliability than their predecessors, discrete transistors and vacuum tubes. The increased reliability of ICs, measured in mean time between failures, is a matter of orders of magnitude, not just marginal improvements. According to one study ICs typically manifest a reliability of about 10^{11} hours per gate;[24] for a chip with 10,000 gates, this reliability figure translates into a mean time between failures of about ten million hours, which is 1,000 years. A discrete transistor will fail on average once every 10^8 hours, and a vacuum tube once in less than 10^6 hours.[25] Furthermore, the failure rate per bit of memory has been declining at the same time that the level of integration has increased. For instance, the failure rate for random-access memories (RAMs) decreased from between 1 and 0.1 failure per bit per billion hours (114 centuries) for 256-bit chips in 1969 to less than 0.01 failure per bit per billion hours for 16K chips in 1977.[26] These enhanced reliabilities were essential for computers to be useful in day-to-day applications. Anyone who has owned a television with vacuum tubes can imagine the horror of tracking down

22. A *bit* is a single piece of binary information, a 0 or a 1. A 64K memory chip therefore can hold 64,000 bits.
23. Franco Malerba, *The Semiconductor Business: The Economics of Rapid Growth and Decline*, 25.
24. A *gate* consists of a transistor and other devices that manage the flow of electricity into and out of the transistor. Several transistors may share, for instance, a capacitor. Thus the reliability for a single discrete transistor would have to be multiplied by the number of devices needed to perform like an IC in order to compare reliability for the whole IC. Reliability for a single IC gate, though, is far higher than for a single transistor.
25. Cited in OTA, *International Competitiveness in Electronics*, 88, n. 26.
26. Ibid., 219–20.

dead tubes on a daily basis in a computer using tens of thousands of them.

The IC has been the vanguard of a technological revolution not only because performance (speed, capacity, reliability) has improved rapidly but also because learning curves and production economies of scale have caused prices to fall. Prices for ICs, for comparative purposes, can be measured by the cost per bit in RAM chips. This measure fell from 1 cent per bit in 1971 (1K RAMs) to 0.05 cents per bit in 1979 (4K RAMs) and to about 0.005 cents per bit in 1985 (64K and 256K RAMs).[27] The combination of steadily improving performance and constantly declining prices has meant that ICs have become increasingly attractive for virtually innumerable applications. Of course, ICs are the basic building block in sectors tightly linked to microelectronics, like consumer electronics (televisions, stereos, videocassette recorders), computers, and telecommunications. But ICs are also finding distant applications in cars, microwave ovens, industrial machinery, toys, and myriad other products.

Growth in Telematics

With the exception of telecommunications, telematics is a young sector. The IC was invented only in 1958, and the electronic computer began commercial production in the early 1950s. Telephones have existed since before the turn of the century, but the advent of microelectronics has so altered telecommunications that it is also virtually a new industry. The telematics industries have enjoyed high growth rates and will continue to increase their share of the gross national product (GNP) over the coming decades.

Semiconductors   The semiconductor business has been a volatile one since the 1960s, with periods of boom and bust. Nevertheless, the overall trend of the industry has been one of explosive growth, as seen in Table 3.2. As the semiconductor market has grown, the share of ICs in the total market has also skyrocketed, as ICs replace discrete devices in more and more applications.[28] In fact, the share of ICs in semiconductor production rose from 58 percent

---

27. Dimitri Ypsilanti, "The Semiconductor Industry," 14.
28. Discrete devices are individual components like transistors, resistors, capacitors, and diodes. ICs comprise thousands of miniaturized devices.

TABLE 3.2.   WORLD SEMICONDUCTOR
MARKET, 1960–86
(in million U.S. dollars)

Year	Market
1960	750
1965	1,700
1970	3,000
1974	5,400
1978	8,910
1982	14,160
1986	26,350

SOURCES: 1960–70: Franco Malerba, *The Semiconductor Business: The Economics of Rapid Growth* (London: Frances Pinter, 1985), 101. 1974: Dmitri Ypsilanti, "The Semiconductor Industry," *OECD Observer* 132 (January 1985):15; 1978–86: Thomas R. Howell, William A. Noellert, Janet H. MacLaughlin, and Alan W. Wolff, *The Microelectronics Race* (Boulder, CO: Westview Press, 1988), 217.
    Notes: Figures for 1960–70 represent consumption. Data for 1978–86 do not include the output of U.S. captive producers and thus somewhat understate total world consumption.

in 1978 to 83 percent in 1985.[29] It is also worth remembering that these market figures are given by value, not volume. This is significant because for many semiconductor products the price per unit was declining rapidly even as total sales soared. For instance, the price of the 64K RAM chip was expected to level out at about $50 in 1980. By 1981–82 it had plunged to $10, then to $5.50–$7 in mid-1982, and on down to $3.50–$5 by the end of the year.[30]

In the early 1980s, the period in which European telematics collaboration took shape, IC markets were expanding. Table 3.3 depicts average annual growth in production and demand for ICs during that period. IC markets were clearly growing much more rapidly than GNPs, which were struggling through the recession following the 1979 oil crisis. The table also shows why Europeans were be-

29. OECD, *The Semiconductor Industry: Trade Related Issues,* 102.
30. Ypsilanti, "Semiconductor Industry," 20.

TABLE 3.3.   AVERAGE ANNUAL GROWTH RATES, IN
PERCENT, FOR IC PRODUCTION AND DEMAND,
1978–82, BY REGION

	United States	Japan	Western Europe	Other
Growth rate in demand	20	19	13	18
Growth rate in production	17	25	12	18

SOURCE: Organization for Economic Cooperation and Development, *The Semiconductor Industry: Trade Related Issues* (Paris, 1985), 103.

coming concerned about their telematics industries: Both production and demand growth (a rough indicator of diffusion of IC technologies) lagged significantly behind other regions.

In addition, semiconductor business has been extremely profitable for the winners. The seven major merchant semiconductor companies in the United States reported after-tax earnings of an (unweighted) average 6.5 percent of sales in 1978 and an (unweighted) average 16.5 percent of equity.[31] Comparable averages for the Fortune 500 industrial firms that year were 4.8 percent and 14.3 percent, respectively.[32]

Finally, and to rely again on American data, the semiconductor industry has experienced growth in employment and productivity since 1970, at least in the United States and Japan. Employment in the microelectronics sector grew in the United States from 115,200 in 1972 to an estimated 230,000 in 1982—an average annual increase of 7.2 percent—despite a massive shift by U.S. semiconductor producers to overseas production facilities (perhaps as much as 90 percent of merchant firms' assembly work).[33] From 1965 to 1980 productivity (measured by value added per production-worker hour) in the semiconductor industry grew faster than productivity in U.S. manufacturing industry generally, surpassing it around 1971.[34]

31. *Merchant* semiconductor houses are those that produce high-volume, commodity (as opposed to custom or semicustom) chips for sale to end users. *Captive* semiconductor production is that carried out by integrated firms mostly for in-house consumption in their own products. For example, IBM is a major producer of ICs but almost entirely for use in its own machines.
32. OTA, *International Competitiveness in Electronics,* 529.
33. Ibid., 137, 350, 357.
34. Ibid., 352.

Computers   Like semiconductors, indeed because of semiconductors, electronic data-processing industries have taken off since
1970. The radical improvements in semiconductor technology, both
in price and performance, made possible similar progress in computers. The first computers were large, unwieldy, power-hungry beasts
whose insides consisted of banks of unreliable vacuum tubes. Not
surprisingly, the market for such machines was thought to be limited. For example, when the U.S. Bureau of the Census purchased
a Univac I (the first computer sold commercially) in 1951, some
people predicted a market for digital computers of perhaps twelve.
When sales to private industry began within a few years, the potential market was estimated at perhaps fifty American corporations.[35] By 1960 the number of computer systems installed in the
United States had reached 5,500; in 1970 there were some 65,000;
and in 1983, an estimated 400,000—and that figure does not include desktop and other very small computers![36] In fact, by 1988,
18 percent of American households owned a personal computer.[37]

What accounts for this striking growth in the production and
consumption of computers? Clearly, semiconductor advances made
possible the growth of the computer industry. More specifically,
falling prices plus the increased speed, capacity, and reliability of
ICs made computers increasingly powerful, reliable, and inexpen-
sive. Take first the sheer practicality of installing a computer after
the advent of ICs. Table 3.4 compares IBM's 650 model from 1955,
a vacuum-tube machine, with Fairchild's F-8 microcomputer from
the 1970s. With improvements in semiconductors computers became smaller, lighter, and less power-hungry. Owning one became
possible for small and medium-size businesses.

The chief factor in the growth of computer markets was, of course,
price. Following the trend in ICs, computer prices have steadily declined. For instance, the PDP-8, a popular minicomputer introduced
by Digital Equipment Corporation (DEC) in the mid-1960s, started
at $18,000 but by the early 1970s some versions could be had for
$2,500.[38] The cost of memory storage has plummeted: The storage
cost for one million data characters in an IBM high-speed disc de-

35. L. M. Branscomb, "Electronics and Computers: An Overview," 755.
36. OTA, *International Competitiveness in Electronics,* 151.
37. "Watch Out, TV—VCRs Are Winning Our Hearts," *San Jose Mercury News,*
24 June 1988, p. D14.
38. OTA, *International Competitiveness in Electronics,* 88.

TABLE 3.4   IBM 650 VS. FAIRCHILD F-8

Characteristic	IBM 650	F-8
Physical volume, in cubic feet	270	0.01
Weight, in pounds	5,650	1
Power consumption, in watts	17,700	2.5
Memory, in bits	3K main, 100K secondary	16K ROM, 8K RAM
Central processor	2,000 vacuum tubes	8-bit microprocessor units plus many discrete devices
Time for adding two numbers, in microseconds	750	150
Reliability, as mean time between failures	Hours	Years
Price	$200,000 (1955 dollars)	Under $1,000 with terminal

SOURCE: Office of Technology Assessment, *International Competitiveness in Electronics,* OTA-ISC-200 (Washington, DC: U.S. Congress, 1983), 88.

vice in 1987 was about one-tenth of the 1976 cost, in constant prices.[39] Plus, declining costs have been accompanied by improved performance. Control Data Corporation's (CDC) 1972 CDC Cyber 176, a high-speed mainframe, could make 9.1 million arithmetic computations per second. The 1981 CDC Cyber 205 could make 800 million such calculations per second.[40]

Before depicting the growth of the world computer market it is necessary to explain the various segments of that market, which can be distinguished on the basis of the size of the machine.

Microcomputers are small machines (they can fit on a desk or, with laptops, on your lap) with one or a few microprocessors. A thirty-two-bit microprocessor is currently standard. Popular microcomputers include the Apple Macintosh family and the IBM PC and PS/2 families and their clones. They are relatively in-

39. Tim Kelly, *The British Computer Industry: Crisis and Development,* 10.
40. OTA, *International Competitiveness in Electronics,* 88.

expensive (from less than $1,000 up to $10,000) and are commonly used in offices, small businesses, schools, and homes.

Minicomputers are about the size of a desk and are commonly used for data-processing in laboratories, factories, and businesses. The microprocessors used in minicomputers work with thirty-two-bit words. A popular mini is made by DEC (the PDP-11), which also makes a widely used supermini (the VAX series). Prices range from $10,000 to $100,000.

Mainframes are large machines that use several thousand high-speed logic chips, not just one or several, and operate on thirty-two- or sixty-four-bit words. They can support several different terminals and peripherals (keyboards and video displays, printers, disc or tape storage units). IBM is by far the largest producer of mainframes, though many other companies build comparable machines. Mainframes can cost from $1 million to $5 million.

Supercomputers, especially large and fast machines, are used for complex scientific calculations (in meteorology, aeronautics research, nuclear research). Only a few firms make them (Cray, CDC, Fujitsu), and they run $10 million and up.

The increasing power of ICs has blurred the lines between types of computers. A top-notch microcomputer today can perform the tasks that used to require a minicomputer. And some superminis are so powerful that they can do the major number crunching formerly reserved for the mainframes. In addition, computer networking (linking several minicomputers or a mainframe with numerous microcomputers or a mini with micros) is emptying the classification of some of its meaning. The key at present is to acquire the right combination of machines and to tie them together as usefully as possible.

Table 3.5 reveals the striking growth in computer markets in the advanced industrialized countries—markets which more than tripled from 1975 to 1982. These figures do not even include the market for software, which was worth some $30 billion worldwide in 1985.[41] Like semiconductors, computers have been extremely profitable (Table 3.6). Although these figures reflect the extraordinarily profitable position of IBM through the years, it is still clear that the American computer industry has been significantly more prof-

41. OTA, *International Competition in Services,* 157.

TABLE 3.5.   WESTERN COMPUTER SALES
(in billion U.S. dollars)

	1975	1982
United States		
Microcomputers and minicomputers	$ 0.86	$13.81
Mainframes	5.40	13.00
Memory and storage	2.95	4.14
Other peripherals	3.62	13.20
Total	$12.83	$44.15
Western Europe		
Microcomputers and minicomputers	$ 0.37	$ 4.54
Mainframes	2.79	8.00
Memory and storage	1.40	*
Other peripherals	1.24	3.58
Total	$ 5.80	$16.12
Japan		
Microcomputers and minicomputers	$ 0.17	$ 3.19
Mainframes	1.89	2.63
Memory and storage	0.44	1.60
Other peripherals	0.53	3.61
Total	$ 3.03	$11.03
Total	$21.66	$71.30

SOURCE: Office of Technology Assessment, *International Competitiveness in Electronics,* OTA-ISC-200 (Washington, DC: U.S. Congress, 1983), 146, 151.
*Included under "Other peripherals"

itable on average than American manufacturing generally. Finally, other computer-related industries, like data-processing services, have not even been mentioned. Because data-processing services depend heavily on telecommunications capabilities, I discuss them in the following section.

Telecommunications   The IC has transformed telecommunications equipment and services. Until the 1970s telecommunications networks operated on the basis of analog circuits—that is, the electronic impulses carrying signals through the system modulated over a continuous range, imitating the original message (for ex-

TABLE 3.6.   RETURN ON EQUITY IN THE U.S.
COMPUTER INDUSTRY, 1974–82, PROFITS AS A
PERCENTAGE OF VALUE OF COMMON STOCK

	Computer Composite[a]	All-Industry Composite
1974	16.9	14.0
1976	17.9	14.0
1978	20.4	15.1
1980	19.2	15.3
1982	15.9	11.0

SOURCE: Office of Technology Assessment, *International Competitiveness in Electronics,* OTA-ISC-200 (Washington, DC: U.S. Congress, 1983), 147.
[a]Includes office equipment and computers

ample, the human voice). Digital ICs allow the telecoms network to operate on the basis of digitized information (strings of "1"s and "0"s), just like computers. In fact, telecoms switches increasingly resemble large computers: They are programmable, digital machines operating on the basis of ICs. Anything that can be converted into digital form can be transmitted through the telecoms network: Voices, computer data, document facsimiles, and video pictures can, in principle, course simultaneously through the emerging Integrated Services Digital Networks (ISDN) in the form of packets of electronic pulses. As ISDN and broadband systems replace the old analog systems, and legions of new services come on line, investments and production in the telecommunications sector are soaring.

At the most general level, in 1985 telecoms equipment and services markets were worth about $115 billion in the United States, or 5 percent of gross domestic product (GDP), about $65 billion in Europe, and $25 billion in Japan (about 3 percent of GDP in Europe and Japan). Furthermore, the telecommunications sector is growing faster than the economies of the industrialized countries and will likely account for 7 to 10 percent of GDP in the West in the 1990s.[42] The world market for telecommunications equipment more than doubled in value (in constant dollars) from 1977 to the

42. Michael Borrus et al., *Telecommunications Development in Comparative Perspective: The New Telecommunications in Europe, Japan and the U.S.,* 2.

mid-1980s, to about $90 billion.[43] Also, output in telecommunications equipment expanded much more rapidly than overall manufacturing output for most OECD countries surveyed during the 1970s. Furthermore, prices for telecommunications equipment (following the trend in ICs) rose more slowly than overall manufacturing prices for the same decade. The OECD therefore concludes that the growth rate in real output for telecoms equipment has substantially outstripped that for manufacturing in general.[44]

Correspondingly, markets for telecommunications services are booming. Telecoms services in 1986 were worth a total of about $64 billion in Europe, $30 billion in Japan, and $116 billion in the United States.[45] Furthermore, advanced services have been growing more quickly than basic services (voice transmission). The data-processing-services market alone reached a total value of $26.4 billion in 1986.[46] In the EC enhanced (or value-added) services have been growing at 25 to 30 percent per year.[47]

Economic Spillovers

Taken together, the telematics sectors (components, computers, telecoms equipment and services) had a total world value in the mid-1980s of some $695 billion.[48] Yet the greatest economic importance of telematics derives from its ripple effects throughout the economy. Telematics comprises a cluster of leading technologies that are driving a technological revolution across the entire economic front. It is beyond the scope of this study exhaustively to catalogue and quantify the economic spillover effects of telematics; I do attempt to paint a verbal picture of the pervasiveness of change induced by telematics technologies throughout the industrialized economies. What follows, then, is illustration, not measurement.

To begin with, ICs are now built into products that have been around for a long time. For instance, electronic watches devastated the producers of mechanical timepieces. Automobiles employ semiconductors to control electronic ignition systems and antilock brake

43. OECD, *Telecommunications: Pressures and Policies for Change,* 21; and Hubert Ungerer, with Nicholas P. Costello, *Telecommunications in Europe,* 93.
44. OECD, *Telecommunications,* 25–26.
45. Ungerer, *Telecommunications in Europe,* 103.
46. OTA, *International Competition in Services,* 183.
47. William Dawkins, "Row Looms after EC Decision on Telecoms Services," *Financial Times,* 15 December 1988, p. 20.
48. Ungerer, *Telecommunications in Europe,* 93.

systems. Instruments of all kinds—scientific, medical, industrial—can now achieve levels of accuracy and speed previously unattainable, in everything from scales to test equipment for quality control to computerized axial-tomography scanners. The latest models of airliners (not to mention military aircraft), like the Airbus A-320, have completely computerized flight controls. The electronic controls replace hydraulic and mechanical connections to the flaps and ailerons, and the computer can override the pilot if he or she attempts a maneuver that would, for instance, cause a stall.

Microelectronics has permeated consumer goods. The brain of the calculator, which has guaranteed that generations of students graduate unacquainted with the slide rule, is a chip. ICs made up 5 to 7 percent of the value of a color television in 1987; that proportion is bound to rise as high-definition TVs (HDTVs), which rely on powerful chips, enter production. The list of consumer goods run by ICs is enormous and growing: stereo systems, microwave ovens, videotape recorders, video games, "talking" dolls, telephones, answering machines, and more.

Instead of looking at specific products, we can also look at whole classes of business activity and note the changes wrought by telematics. Take retailing: Electronic cash registers can now be connected to computers that keep track of sales (by product) and inventory. Or wholesaling and distributing: Customers can send their orders electronically to the distributor's central computer, which then allocates orders to various warehouses and plans the optimal loading and routing of trucks.[49] Publishing is no longer a matter of typing and typesetting. For instance, newspaper reporters can now compose their stories on small (even laptop) computers. Their articles then feed electronically into a larger computer on which the editors arrange the page layouts. The computer then transmits the electronic pages to a machine that prepares the actual "plates" (they are not sheets of metal type but more like a template). Even further, newspapers (like the *New York Times* and the *Wall Street Journal*) can publish West Coast editions by beaming electronic data via satellite to printing plants in California.

The banking system is now thoroughly computerized; new services like automated tellers and home banking link customers di-

49. See the accounts of business users of telematics in François Bar and Michael G. Borrus, *From Public Access to Private Connections: Network Policy and National Advantage.*

rectly to a bank's computer, and transactions take place electronically. Computers have enhanced the ability of financial analysts
and traders to track stocks and other financial instruments. Trading
in stocks, bonds, securities, futures, and currencies can take place
almost instantaneously and across the globe. The Wall Street crash
of October 1987 called into question the power of computers over
markets.

Manufacturing industry is also being revolutionized by telematics. Rapid, graphics-oriented computers are replacing the drawing
board for designers and engineers. Computing power has transformed machine tools; these can now be quickly reprogrammed to
produce small batches of differentiated parts. The result is an
emerging new structure for industry, as small machine shops can
cheaply manufacture custom parts on small runs.[50] Telematics is
also transforming large-scale manufacturing. Whole production lines
can now be automated by computer control and linked to suppliers,
warehouses, and customers to optimize the flow of inputs and outputs. Or, as Stephen Cohen and John Zysman point out, "programmable automation" offers the possibility of manufacturing
several different products in small batches on large production lines.[51]

The new telecommunications permits new means of managing
businesses at geographically far-flung sites. Design and production
specifications, even blueprints, can be transmitted electronically,
eliminating the need for expensive travel and in-person consultation. Production and shipments can be monitored from across the
world, and business meetings can take place through videoconferencing. As François Bar and Michael Borrus put it, "Companies are
discovering they can use the new telecommunications technologies
to improve their operations and modify their competitive environment to their advantage."[52]

SUMMARY

Technological revolutions are thoroughgoing transformations of the
economy based on clusters of radical and incremental innovations.
Such revolutions, like the mid-nineteenth-century impacts of steam

50. See Michael J. Piore and Charles F. Sabel, *The Second Industrial Divide:
Possibilities for Prosperity.*
51. Cohen and Zysman, *Manufacturing Matters,* chaps. 8–11.
52. Bar and Borrus, *From Public Access,* 4.

power and iron, alter the industrial landscape by sending some sectors into decline and building up new poles of growth. They thus alter the terms of economic growth and competition. Telematics is at the core of a technological revolution in the industrialized countries that is creating large new industries and transforming most traditional ones. I argued that state policies can make a difference in the extent to which nations adapt to and benefit from technological revolutions, with Japan as the example of a nation that has used policy intervention successfully.

However, not all governments will institute policy measures to derive advantage from technological change. The United States, in fact, has not moved far in the direction of technology or industrial policy; public involvement in high technologies in the United States flows chiefly through the Pentagon. But we would expect the Western European countries to make the telematics revolution a focus of public policy. The basis of that expectation lies in the deep-rooted tradition of intervention in industrial practices in postwar Europe. Industrial policies are a fixture in Europe, even under governments ideologically inclined to favor market mechanisms. Telematics became too important economically to be allowed to lag and founder; it became for European governments a vital arena for intervention. That is the story I take up in the next chapter.

FOUR

National Telematics Policies in Europe

Not surprisingly, the technological revolution emanating from telematics grabbed the attention of European political elites committed to economic growth. As the economic importance of telematics became increasingly clear to European governments, the perceived costs of missing out on the benefits of the telematics revolution soared. Ultimately, the Europeans decided that they could not afford to cede telematics to the United States and Japan, even if it meant sacrificing some autonomy to collaborate.

Collaboration, though, is the denouement of this story. We must begin by presenting the dramatis personae and setting up the tension. Arriving at collaboration was no simple matter because each of the major European states (the Federal Republic of Germany, France, Italy, and the United Kingdom) had made telematics a national priority and built up national-champion companies to carry their flag in domestic and international competition. This chapter portrays the rise of telematics policies in Europe and describes the national-champion strategies that emerged in all the major countries.

I begin the chapter by summarizing the political concern for "big science" and science policy in postwar Europe. Military objectives were important (especially for France and Great Britain), and policy initially focused on scientific research. In time the economic value of science and technology rose in prominence, with a shift in policy emphasis from science to technology in the 1960s. Policy-makers eventually, in the 1970s, linked technology to industrial-policy concerns. As telematics revealed its importance as a leading sector for the economy as a whole, Europe's welfare democracies were motivated to find ways of building up telematics industries.

The chapter then narrows in focus to examine telematics policies in Europe. My starting point is the technology-gap scare of the late 1960s. European fears of losing out on emerging high-growth, high-value-added industries provoked a series of national-champion strategies in the larger states, first in computers and then in micro-electronics. These strategies aimed at promoting an industry within national borders (frequently even a single firm) that would be competitive domestically and eventually internationally. I also review national telecommunications policies in Europe, which have been highly nationalistic.

SCIENCE AND THE STATE IN EUROPE

During the first half of this century, leaders of states discovered that the R&D efforts of scientists could be channeled in ways that redounded to the increased power, wealth, and prestige of the nation. Many of the crucial experiences in this regard derived from war and the attendant pressures to gain an edge in technologies of destruction. For instance, state support for aeronautical R&D, begun in earnest after World War I, offered a glimpse of the future of war conducted from the air.

The Second World War, though, proved once and for all that governments could marshal scientific and engineering resources with spectacular, even decisive, results. Wartime laboratories produced stunning advances in radar, rocketry, and, of course, atomic energy. Indeed, the Manhattan Project flashed the message across the globe that science mobilized by the state could drastically cut the time between basic research and application, and could reshape the very nature of state power and international politics. As Jean-Jacques Salomon has noted,

> During the second world war, scientific research was used for the first time as a source of new technologies whose influence was to be no less decisive on the post-war period than on the termination of hostilities. After that, political power could no longer leave science to itself but, on the contrary, had to force the pace of discovery and innovation.[1]

1. Jean-Jacques Salomon, *Science and Politics*, 48.

Initially, at least, state-driven science was tied to military ends, especially in those countries harboring ambitions for independent military power, like the United States, the Soviet Union, France, and the United Kingdom. The sectors first chosen to receive massive infusions of public monies were therefore those most closely linked to military needs. Nuclear energy and aerospace were the big-ticket R&D areas in the 1950s and the 1960s, and civil objectives were inextricably bound up in military ends. An OECD study of government-sponsored R&D notes:

> The first government R&D programmes in these fields [space and nuclear energy] were initiated mainly for military purposes and when civil work in the pioneer countries followed, it was usually undertaken by the same organisation as military R&D, drew on a common programme of underlying research and was often voted funds under the same budget heading. It was, therefore, difficult to break space and nuclear R&D down between 'military' and 'civil.' This was sometimes not even attempted by national authorities.[2]

During the 1960s countries with elevated military aspirations invested heavily in expensive nuclear and aerospace R&D areas. Government R&D expenditures thus accounted for the largest share of GNP in those countries pursuing military aerospace goals (Sweden) or military aerospace plus nuclear energy goals (the United States, France, the United Kingdom); see Table 4.1.

During the 1950s and early 1960s science policy in Europe meant state support for basic research. R&D resources (money, personnel, facilities) can be applied along a continuum from fundamental scientific research at one end to commercial development at the other. The objective at the basic-research extreme is the advancement of scientific knowledge, with no regard for possible economic payoffs. At the other extreme, R&D money and personnel are devoted to developing specific products and the processes for making them, with a view directly to the market. In between fall applied research and industrial R&D. The continuum looks like this:

Basic Research	Applied Research	Industrial R&D	Commercial Development

In Europe up to the mid-1960s governments supported R&D for basic research and some applied research. To the extent that links

2. OECD, *Changing Priorities for Government R&D*, 81.

TABLE 4.1.   GOVERNMENT R&D FUNDING AS A
PERCENTAGE OF GNP, 1961–70

	1961	1964	1966	1968	1970
Belgium	—	—	0.6	0.6	0.7
France	0.9	1.2	1.4	1.4	1.2
Germany	0.5	0.8	0.9	0.9	0.9
Italy	—	0.2	0.3	0.4	0.5
Japan	0.6	0.4	0.5	0.5	—
Netherlands	0.5	0.7	0.9	0.9	0.9
Sweden	0.8	1.0	1.0	1.1	—
United Kingdom	1.4	1.4	1.3	1.2	1.2
United States	2.0	2.3	2.2	1.9	1.6

SOURCE: Organization for Economic Cooperation and Development, *Changing Priorities for Government R&D* (Paris, 1975), 118–19.

to economic development were pondered, policy-makers assumed that fundamental scientific discoveries and breakthroughs would flow into economically useful applications by a sort of natural process (the technology-push perspective mentioned in Chapter 3).

For instance, in the United Kingdom the postwar Advisory Council on Scientific Policy concerned itself almost purely with basic research; its 1961–62 annual report said nothing about technology.[3] Government R&D spending was uncoordinated, with 70 percent of public R&D expenditures flowing through the various mission-oriented departments. The remaining 30 percent was lumped in a category called "universities and scientific research."[4]

From Science Policy to Technology Policy

In the mid-1960s began a shift to what I shall call technology policy, meaning state measures to promote R&D and the diffusion of scientific advances and technical innovations in industry. Technology policies in Europe moved government intervention on the research continuum in the direction of industrial R&D. In other words, policy elites in Europe began to link science policy to economic

3. Freeman, *Technology Policy*, 120–21.
4. OECD, *Changing Priorities*, 26.

objectives. Indeed, an article in *Science* magazine in October 1969 noted that "science policy-making in Western Europe has entered a utilitarian period. . . . The reason, quite simply, is that the major industrial nations, with the exception of Italy, . . . are eager to emulate the American pattern of close ties between research and industry."[5] A highly revealing way of tracking the changes is to follow the organizational shake-ups among the bureaucracies charged with science, technology, and industry in Europe.

In Britain the new Labour government of Harold Wilson, determined to reforge Britain's economy in the "white heat of technology," created a Ministry of Technology in 1964. It also passed the Science and Technology and the Development of Innovations Acts in 1965, both of which increased the amounts of government R&D monies for industry.[6] The Research Councils (for medicine, agriculture, science, and natural resources) were hived off to the new Department for Education and Science.[7] In 1969 the Science Research Council (SRC) announced a realignment of its objectives, to place an emphasis on "research and training related to industrial needs." The SRC had already begun using its funds to encourage "more scientific talent to enter industrial research."[8] The Conservative government that came into power in 1970 merged the Ministry of Technology with the Department of Economic Affairs and the Board of Trade, creating the Department of Trade and Industry (DTI). That move symbolized better than anything else the marriage of science and technology with economic objectives. In the mid-1970s the Advisory Board for the Research Councils and the Advisory Council on the Application of Research and Development (ACARD) led "an increased effort to make the country's scientific and technical resources more responsive to the needs of the economy and of social policy."[9]

In France a similar evolution took place. Under Charles de Gaulle, of course, scientific preeminence was a national goal, and the state supported scientific research oriented to prestige and military independence. For instance, in 1963, the appropriation for basic research in the national laboratories and universities totaled only 15

5. D.S. Greenberg, "Britain: New Emphasis on Industrial Research," 485.
6. Geoffrey Shepherd, "United Kingdom: Resistance to Change," 160.
7. Freeman, *Technology Policy,* 121.
8. Greenberg, "Britain: New Emphasis," 485.
9. Freeman, *Technology Policy,* 122.

percent of the total government expenditure on R&D, with defense and nuclear energy absorbing the bulk of the money.[10] The framework for creating science policy was set up in 1958, with an interministerial committee to coordinate research budgets of the various bureaucracies; an advisory committee of prominent scientists and technologists, à titre personnel (chosen on individual merit, not on institutional affiliation); and a secretariat for the government's science efforts, the Délégation Générale à la Recherche Scientifique et Technique (DGRST).

The shift toward economic objectives for R&D policy began during the formulation of the Fifth Plan in 1964 and 1965 (the Plan would cover 1966–70). Reportedly, the Commissariat du Plan (charged with economic planning) and the DGRST worked more closely together than ever before to elaborate the portion of the Plan on science. During the 1967 legislative elections the minister in charge of science, Alain Peyrefitte, argued that French industry should receive government assistance to invest in the R&D needed to close the gap with American high-tech industries. One step taken by the government that year was to create the Agence Nationale pour la Valorisation de la Recherche (ANVAR), whose mission would be to promote contacts between researchers in industry and the universities.[11] ANVAR later acquired the purpose of aiding small companies to adopt and exploit new technologies.

On taking office in 1969, Georges Pompidou's government reshuffled ministries as the British would the following year. The Ministry of Science and the Ministry of Industry were combined to form the Ministry for Industry and Scientific Development. The new ministry held cabinet rank, which the previous Ministry of Science did not. Its mandate was to increase the benefits to industry of massive public investments in research. Also significantly, the first minister for industry and scientific development was an economist, François-Xavier Ortoli.[12] As in Britain the melding of the science and industry ministries was the perfect institutional expression of a shift from science policy to technology policy.

The movement of policy along the continuum from basic research toward applied and industrial R&D was not as striking in

10. John Walsh, "Some New Targets Defined for French Science Policy," 628.
11. Ibid., 629.
12. D.S. Greenberg, "France: Profit Rather Than Prestige Is New Policy for Research," 1335–36; Greenberg, "Britain: New Emphasis," 485.

Germany and Italy as it was in France and Britain. Nevertheless, it occurred. Germany established a Ministry of Science in 1962; by the early 1970s it also handled federal education programs. The Federal Ministry for Research and Technology (BMFT) was split off from the Ministry for Education and Science. The Ministry of Education and Science was to oversee research in the universities, while the new ministry was given the mandate to administer federal research policies in specific fields like data-processing, aerospace, nuclear energy, and marine sciences.[13] In time the role of the BMFT became to promote "the development of key technologies and help restructure and modernize the economy."[14] Thus, the BMFT has controlled programs and specific projects for industrial R&D. A complicating factor in Germany is that the *Länder* have prime authority over higher education (university research) and also support R&D in industry and the independent institutes (like the Max-Planck Society and the German Research Association).

Italy, alone among major European countries in 1970, had no cabinet-level ministry of science (though it had a Minister for Coordinating Scientific and Technical Research, without portfolio).[15] The key body for science in Italy was for a long time the National Research Council (CNR), which prepared the annual research budget. Even the CNR held limited powers, though, as it administered (around 1970) only about 20 percent of government R&D funding.[16] Science policy was continuously in a state of chaos.[17] Nevertheless, it did move toward support for industrial technologies as elsewhere in Europe. The first industrial R&D program emerged in 1968, when parliament created the Fondo per la Ricerca Applicata within the state financial institution (though it was subsequently attached to the new Ministry of Scientific Research). The Fondo has the authority to confer easy credits worth up to 90 percent of the costs of R&D projects in private firms.[18]

In short, by 1970 all the major European countries had shifted their science policies away from an almost exclusive focus on state-dominated sectors like defense and nuclear energy. The movement

13. OECD, *Changing Priorities*, 41–42.
14. Ernst-Jürgen Horn, "Germany: A Market-Led process," 51.
15. D.S. Greenberg, "Science in Italy: Reform Effort Takes a Sharp Turn Leftward," 1706.
16. OECD, *Changing Priorities*, 44–46.
17. D.S. Greenberg, "Italy: OECD Report Finally Emerges," 587.
18. Pippo Ranci, "Italy: The Weak State," 139.

TABLE 4.2.   GOVERNMENT R&D FUNDING BY
MAJOR CATEGORIES AS A PERCENTAGE OF GNP,
1961–69

	National Security and Big Science[a]			Economic Development[b]		
	1961	*1965*	*1969*	*1961*	*1965*	*1969*
Belgium	—	0.13	0.17	—	0.12	0.17
France	0.62	0.93	0.68	0.07	0.14	0.25
Germany	0.19	0.31	0.39	—	—	0.09
Italy	—	0.14	0.17	—	0.00	0.03
Japan	0.05	0.02	0.05	0.13	0.14	0.13
Netherlands	0.09	0.16	0.19	0.11	0.15	0.22
Sweden	0.56	0.61	0.37	0.07	0.10	0.14
United Kingdom	1.10	0.93	0.67	0.14	0.15	0.30
United States	1.73	1.93	1.33	0.06	0.06	0.10

SOURCE: Organization for Economic Cooperation and Development, *Changing Priorities for Government R&D* (Paris, 1975), 80, 127.
[a]Defense, civil nuclear, civil space
[b]Agriculture, mining and manufacturing, economic services

was toward technology policies, reflecting the growing concern for applications of science and technology to industry. The OECD charts this process with figures comparing the share of GNP taken by government R&D spending for broad categories of objectives (Table 4.2). The key trend to note is the rising share of "economic development" everywhere through the 1960s.

From Technology Policy to Telematics Policies

From a concern for the economic payoffs of scientific and technological research it was a short step to concern for the industries based on such research. Technology policy began to meld with industrial policy in the mid-1960s in Western Europe. By *industrial policy* I mean the efforts of governments to influence the allocation of resources among or within industrial sectors. Industrial policies reflect the conviction that "normal" market processes do not produce the socially optimal mix of industries.[19] As Tyson and Zysman point out, the tools of industrial policy vary over a broad spectrum. They range from the macroeconomic (fiscal and monetary policies) to market-promotion policies (aimed at improving the functioning of markets for labor, capital, and products) to sector-specific measures (designed to meet the needs of individual sectors or firms).[20]

All the countries of Western Europe have implemented industrial policies of various kinds since World War II. The government organs created to administer Marshall Plan funds carried out these policies. Postwar governments in Europe sought to rebuild as quickly as possible the basic industries (steel, coal, cement) needed to resume growth. The allocation of credits among industrial activities constituted a sector-specific approach to reindustrialization. Afterward, none of the European governments wholly renounced measures designed to shape the industrial mix of their economies. The French, through their Commissariat du Plan, adopted detailed sector-specific industrial policies. The Germans, though ideologically inclined to rely on macroeconomic or market-promoting industrial policies, also intervened in specific sectors, like coal, aerospace, and, as we shall see, telematics. The British have bounced around between the extremes represented by France and Germany, flirting with but never attempting full industrial planning, and implementing sector-specific policies while proclaiming allegiance to economic liberalism (nonintervention).

Once the European economies were on solid footing, the social commitments of European governments acted to ensure that industrial policies would continue to be an important part of the landscape. The European welfare democracies established social bargains that included two main planks: redistribution of national income in favor of the economically weak (the poor, unemployed, and retired), and protection of employment. Thus, European governments have spent a higher share of the national product than have American or Japanese governments.[21] A prime objective of popularly elected governments in Europe therefore became the economic growth needed to pay for social programs and maintain em-

19. This conception of industrial policy coincides with that of Alexis Jacquemin, "Introduction: Which Policy for Industry?", 1.
20. Tyson and Zysman, "American Industry in International Competition," 19–22.
21. François Duchêne and Geoffrey Shepherd, *Managing Industrial Change in Western Europe*, 27.

ployment. Attaining growth and full employment justified industrial policies even in those (British and German) governments opposed
in principle to public interference in the market.

Industrial policies under these circumstances have two major kinds
of rationales. First, industrial policies may be necessary for the defense of employment. As Jacques Lesourne argues, where social
groups are organized into powerful associations ("social oligopo-
lies"), governments interested in staying in power must make
concessions to them even at the cost of economic inefficiency. For
example, governments subsidize inefficient or uncompetitive industries in order to maintain employment, thus buying peace with powerful labor associations. As Lesourne notes, arguments for industrial policies in European countries with strong social organizations
"are based above all on the defence of employment. Hence the fact
that present European industrial policies often have a strong back-
ward-looking component."[22] The second major rationale for industrial policies is based on the notion, which gained currency during the 1960s, that certain industries and sectors are economically
strategic and that a nation must maintain a presence in these sectors
or at least "keep the technical ability to regenerate on its own soil
any 'strategic' industrial activity."[23]

Two trends led to industrial policies favoring strategic sectors.
First, policy-makers came to believe that certain industries were
leading (as defined in Chapter 3). Leading industries drove economic growth across a broad front and thus merited government
support so as to ensure the country's economic future. Already in
1967–68 the OECD's *Gaps in Technology* reports identified computers and electronic components as leading industries.

Second, during the 1960s and 1970s, the European economies
became increasingly involved in an open world economy, as the
various GATT rounds reduced barriers to trade. Under these conditions the newly industrializing countries started to capture export
markets in mature sectors like textiles, shipbuilding, and steel. During the economically expansive years leading up to 1973, employment losses in these sectors were offset by employment gains in the
growing economy as a whole. The costs of defending employment
were not apparent as long as the European economies were growing

22. Jacques Lesourne, "The Changing Context of Industrial Policy: External and
Internal Developments," 28–29.
23. Ibid., 31.

steadily. But after 1973 enduring recession and persistent unemployment made adjustment not nearly so painless.

One way out of the mess appeared to be via economically leading sectors that would stimulate economic growth and so generate new employment. Thus, the adjustment crisis of the 1970s provoked renewed attention to high-growth, high-value-added, high-technology sectors. Within the OECD, debate over policies and prospects for a way out of the adjustment crisis emerged during the late 1970s. The European countries emphasized industrial policies designed to promote growth in leading sectors. In Britain the byword was *positive adjustment,* in Germany *Strukturpolitik,* in France *modernisation.* In every case, telematics was at the core of European strategies for economic renaissance.

By the 1980s, therefore, technology policy had been tightly linked to industrial policies favoring leading sectors, especially telematics. The OECD noted in 1985 that member states' science and technology policies increasingly focused on industrial-policy objectives:

> It is also clear that R&D priorities for most Members are strongly focused on advancing and applying new technologies, usually in some form of relationship with industry, and often directed to specific industrial sectors. This provides yet another illustration of the growing integration of science, technology and industrial policies. The priority R&D areas, especially those centering on industrial technologies, are remarkably similar for many of the countries.[24]

The OECD report also makes clear that the concern underlying these technology/industrial policies is international competitiveness. Broadly speaking, the common fear is that technologically "lagging countries may benefit less from the new technologies and become dependent on the leading countries for their supply."[25]

R&D has remained a high priority for OECD governments, as evidenced by the increasing fraction of public expenditure going to support R&D during a period (early 1980s) of generalized budget austerity. Public funding of R&D has shifted in recent years even further in the direction of economic-development goals (as opposed to other objectives, like the environment, energy, health and social services). The shift toward economic development has been particularly noteworthy in France, Germany, Italy, the Netherlands, and

24. OECD, *Science and Technology Policy Outlook, 1985,* 25.
25. Ibid., 67.

the United Kingdom.[26] In fact, a new term has been coined to des-
ignate the spate of policies linking technology and industry: *innovation policy.*

The notion of innovation policy embraces a wide range of activities and policy instruments. As Roy Rothwell defines it, innovation
policy "includes the whole sequence of activities involved in the
commercial introduction of a new product or process, from basic
research through to marketing and sales."[27] The policy tools include
education and training, information services (databases, libraries,
consultancy services), financial assistance (loans, credits, guarantees), R&D subsidies, government procurement, trade policy, competition and antitrust policies, taxation (depreciation allowances for
equipment, tax exemptions for R&D activities), legal and regulatory arrangements (patents, royalties, environmental protection), and
regional development policies.[28] In other words, innovation policy
includes anything having to do with pushing advanced technologies
into industrial exploitation. The OECD also recognizes innovation
policy as a new category and has begun a series of studies on national innovation policies.[29]

In short, technology policy and industrial policy began to be linked
in the mid-1960s. At the same time, a panicky debate erupted in
Europe over technology gaps that left European industries dangerously behind their American competitors. The policy response was
a set of national-champion strategies in IT (embracing computers
and microelectronics).

TECHNOLOGY GAPS

Western Europe has suffered through two bouts of technology-gap
fever. The first occurred in the mid-1960s, the second in the early
1980s. The earlier crisis proved to be a watershed in two ways.
First, two key beliefs crystallized among European national policy-
makers during the mid-1960s: that IT was crucial for continued
economic development and that Europe's capabilities in IT were
inadequate and threatened by foreign competitors. These key beliefs
have not changed significantly since then. Indeed, the technology-

26. Ibid., 18–19.
27. Rothwell, "Reindustrialisation, Innovation and Public Policy," 68.
28. See ibid., 69.
29. OECD, *Innovation Policy: France.*

gap crisis of the 1980s flared up precisely because Europeans were more convinced than ever that IT (plus telecoms) was important and that foreign competition imperiled Europe's prospects in these crucial sectors.

Second, the technology-gap crisis of the 1960s accelerated and justified the movement of European governments toward creating and sustaining national champions in semiconductors and computers. A national champion is a firm designated by the government to be the nation's major (if not only) producer of a given set of goods. The national champion, according to the idealized script of its public sponsors, will supply most of the needs of the domestic market and will use that base to capture significant export markets in competition with foreign multinationals. Only when these national-champion strategies had failed to achieve their goals could collaborative initiatives overcome the states' desire for autonomy.

The OECD's *Gaps in Technology*

Ministers of science and technology in the OECD countries asked the organization in January 1966 to study the links between science and the economy. Member states wanted an understanding of technologically advanced industrial sectors and the effects of government policies designed to stimulate them. In particular, the OECD governments sought to specify the consequences for national economies of gaps between countries in high-tech sectors. The Committee for Science Policy oversaw the production of six sector studies and a general report. These documents defined the gaps and analyzed their causes and consequences.

The *General Report* concluded that there were indeed technology gaps separating the United States from all the other OECD countries. The United States spent more on R&D: 3.4 percent of GNP as opposed to 1.5 percent of GNP in OECD Europe. American businesses invested more in R&D than did their European counterparts by more than three to one. Further, U.S. R&D included a far higher percentage of expenditure on large-scale programs: "Sixty-three percent of industrial R and D in the United States is on programmes whose total R and D expenditure is more than $100 million per annum. No firm in any European country has an R and D programme of this magnitude."[30] The report as-

30. OECD, *Gaps in Technology: General Report,* 13.

serted that because of thresholds of expenditure below which R&D programs were probably ineffective, "the dispersion of Europe's relatively small efforts deserves attention."[31] Another major disparity was that the U.S. government spent about eight times as much public money on R&D as did the European Economic Community (EEC) taken as a whole. Although a major chunk (56 percent in 1963–64) of this government spending went to defense, space, and nuclear programs, the report noted that the "indirect commercial effects have been considerable."[32]

The upshot was that American firms accounted for about 60 percent of the 140 significant innovations identified by the OECD, and raked in between 50 and 60 percent of the OECD total of payments for patents, licenses, and know-how. Furthermore, the United States diffused innovations more rapidly than Europe into its economy.[33] Although innovations appeared to spread rapidly to Europe from the United States, this diffusion took place increasingly through American direct investment in Europe. In fact, the capital stock of U.S. firms in EEC Europe increased by 450 percent between 1957 and 1966. The report noted several key problems for Europe, including the fragmentation of the European market into national pieces that were too small; the inadequate size of European firms, which prevented them from achieving economies of scale in R&D and production; and the lack of government sponsorship for industrial R&D.

To answer the question Why does this matter?, the OECD report declared that certain industries were vital to the "diffusion of new products and processes throughout the economy." Therefore, growth and improved productivity depended on a country's having access to emerging innovations, preferably within its own frontiers.[34]

In a final section on policy implications the OECD suggested that national governments needed to help industry enhance its ability to originate and diffuse innovations. It also argued that in Europe member countries needed to "develop more effective forms of cooperation in order to overcome the existing fragmentation of markets, industries and technological efforts."[35] To address the problem of scale, concentration of technological resources would have

31. Ibid.
32. Ibid., 14.
33. Ibid., 13–15.
34. Ibid., 30.
35. Ibid., 33.

to occur at the European rather than the national level. And this step would require first of all the development of institutional arrangements to manage European cooperation.

Two of the six sector reports in the *Gaps in Technology* series dealt with information technologies: electronic components and electronic computers. Regarding components, the group of experts concluded that the United States enjoyed a distinct lead over other OECD countries. This lead could be measured by the preponderance of U.S. firms among licensors of technology, the American origins of new technologies and major inventions, and market shares. U.S. space and defense programs probably helped push the new technologies through the early stages of development faster than would have occurred otherwise. But the chief factor in the components gap, according to the OECD, was the large, sophisticated home market enjoyed by American makers.

The report declared that the relatively marginal contribution of European firms to components development might be due to "virtually non-existent" technical communications within Europe and "poor" communications within countries.[36] Although the report downplayed a government role in the area of international cooperation among firms as "impractical," it mentioned that governments could support the creation of a Europe-wide market to overcome "fragmentation" and "economic nationalism."[37]

Technology gaps in computers also had the United States on the leading end. IBM held about 75 percent by value of all installed computers in the West. The seven other large U.S. makers brought the American share of installed equipment to 90 percent.[38] In the way of solutions, the OECD report suggested that governments needed national programs to reach the scale and quality of R&D that were needed and to retain trained personnel. Governments could also encourage the utilization of computers across society. Fragmentation of the European market prevented European makers from benefiting from its potential size and density.

The American Challenge

Jean-Jacques Servan-Schreiber's views in *The American Challenge* agreed in essence with the *Gaps in Technology* reports, though his

36. OECD, *Gaps in Technology: Electronic Components*, 46.
37. Ibid., 126.
38. OECD, *Gaps in Technology: Electronic Computers*, 8.

flamboyant language conveyed a sense of panic that the OECD's documents did not. Here, for instance, is his description of the stakes:

> These figures are important to keep in mind, for electronics is not an ordinary industry: it is the base upon which the next stage of industrial development depends. . . .
>
> A country which has to buy most of its electronic equipment abroad will be in a condition of inferiority similar to that of nations in the last century which were incapable of industrializing. . . . If Europe continues to lag behind in electronics she could cease to be included among the advanced areas of civilization within a single generation.
>
> America today still resembles Europe—with a 15-year head start. She belongs to the same industrial society. But in 1980 the United States will have entered another world, and if we fail to catch up, the *Americans will have a monopoly on know-how, science, and power.*[39]

Servan-Schreiber was particularly adamant about the crucial role to be played by the electronics industries (especially computers): "Development of the electronics industry controls productivity and the modernization of the whole framework of industry and services."[40]

Servan-Schreiber was committed to the need for action at the European level; not even the larger countries (France, Germany, the United Kingdom) could succeed alone. "Only on a Europe-wide level, rather than a national one, could we hope to meet the American challenge on all major fronts."[41] For the crucial computer sector Servan-Schreiber argued that this was "precisely the choice: either an all-European *plan-calcul,* with all the reconversions it will require, or domination by IBM."[42]

Though some of his more hysterical prophecies did not come to pass, Servan-Schreiber's diagnosis of the problems was widely accepted and, in fact, matched in essence that of the OECD. The key factors were: the smallness of European firms, insufficient levels of R&D spending, U.S. government supports for high-tech industries, and European national markets of insufficient size and sophistication.

THE RISE OF NATIONAL CHAMPIONS

In this section I briefly describe national strategies in IT and in telecommunications, treating each sector separately. The national-

39. Jean-Jacques Servan-Schreiber, *The American Challenge*, 13, 101. Emphasis in original.
40. Ibid., 166.
41. Ibid., 111.
42. Ibid., 167.

TABLE 4.3.   COMPUTER MARKET SHARES IN
EUROPE, IN PERCENT, 1967

	Of the Number of Installed Computers	Of the Value of Installed Computers
France		
IBM	43	63
Bull/GE	31	20
Germany		
IBM	55	68
Siemens	4	9
Zuse	8	—
Telefunken	1	—
Italy		
IBM	—	66
GE-Olivetti	—	19
United Kingdom		
ICL	42	45
IBM	29	39
NCR-Elliott	13	4

SOURCE: Organization for Economic Cooperation and Development, *Gaps in Technology: Electronic Computers* (Paris, 1969), 158–67.

champion strategies generally included consolidation of several small companies into one national standard-bearer, R&D subsidies, government procurement preferences for national firms, and functional standards that acted as a barrier to outside companies. Such policies prevailed into the 1980s, when they began to be supplemented by collaborative efforts.

Computers

Telematics policies in Europe began with computers. Since its beginning, the story of the European computer industry has been one of trying to nibble away at IBM's dominant position. In fact, IBM set up research and manufacturing facilities in Europe at an early stage and quickly grabbed a majority of the national markets, as Table 4.3 shows.

France   French support for the computer industry was born in the heyday of de Gaulle's drive for national autonomy in security and economic affairs. Two key events led to the first IT policies in France. The first was the 1964 purchase by GE of a 50 percent stake in France's main computer manufacturer, Machines Bull. The purchase was highly upsetting to de Gaulle. The second event was the U.S. denial of export licenses for powerful IBM and CDC mainframe computers ordered by France's Commissariat à l'Energie Atomique, which was even more galling to the French president. Then GE-Bull decided to drop two Bull machines from its product line, provoking concerns in France over layoffs and the effects on Bull's R&D capacities.[43]

France responded by creating in the summer of 1966 the office of *délégué à l'informatique,* answering directly to the prime minister. The *délégué* was given administrative responsibility for the *Plan calcul,* a program designed to "bring the French computer industry to within competitive distance of American firms."[44] The *Plan calcul* was launched with a budget of FFr 450 million ($90 million at the time) for 1967–70,[45] though it ended up spending about FFr 640 million. It was followed by the second *Plan calcul* (1971–75, FFr 1,030 million) and the third *Plan calcul* (1976–80, FFr 1,438 million).[46]

Another of the *délégué*'s first tasks was to oversee the first restructuring effort for the French computer industry. This reorganization entailed the creation of the Compagnie Internationale d'Informatique (CII) by the government-arranged marriage of two smaller firms (Société d'Electronique et d'Automatique and Compagnie Européenne d'Automatique) in 1967. Later, in 1975, France torpedoed the only European collaborative initiative in computers (Unidata) by fostering the merger of CII (which was partially state owned) and Honeywell-Bull, with the government taking a majority position in the new company. CII-Honeywell-Bull (CII-HB) received FFr 1,326 million from the third *Plan calcul,* more than nine-tenths of the total for the program. The new national champion also benefited from additional funds to cover the losses of CII over previous years and an assured FFr 4 billion in government purchases over the period 1976–80.[47]

43. John Walsh, "France: First the Bomb, Then the 'Plan Calcul,' " 767–69.
44. Ibid., 767.
45. Ibid., 770; Giovanni Dosi, "Institutions and Markets in High Technology: Government Support for Microelectronics in Europe," 186.
46. Kenneth Flamm, *Targeting the Computer: Government Support and International Competition,* 155.
47. Ibid., 156.

Federal Republic of Germany   During the technology-gap crisis, Germany increased federal spending on science and technology. Several priority sectors received particularly generous increases, one of them being data-processing. The first electronic data-processing program, from 1967 to 1970, was budgeted at DM 387 million, with funds allocated to computer R&D, promoting computer applications, basic research, and education. The second program received DM 2.41 billion (more than six times the budget of the first program) to cover the period 1971–75. The funds went for R&D in computers, peripherals, software, applications, and components, though about half the monies were channeled to the universities and research institutes for basic research and education. The third program (1976–79) maintained the level of funding for industrial R&D but reduced expenditures for education. It distributed DM 1.58 billion.[48]

The German government did not have to put together a national champion in computers because one already existed. Siemens was by far Germany's largest producer in all the telematics sectors— semiconductors, computers, and telecommunications. In fact, Jeffrey Hart shows that BMFT funding for computer and semiconductor R&D in the late 1970s was highly concentrated: Out of a total of $1.4 billion, Siemens received $1.3 billion.[49] Ironically, Germany's greatest success story in computers was Nixdorf, a company launched in the 1950s. Nixdorf shunned government aid and has not relied on its home market—less than half its sales occur in Germany, and 20 percent come from the United States. Nixdorf in the early 1980s held 10.5 percent of the European market for small business systems, in direct competition with successful American firms like DEC.[50]

United Kingdom   Along with the creation of the Ministry of Technology in 1964, Wilson's Labour government also launched its Advanced Computer Technology Project. The first government subsidies for the sector were research grants totaling £5 million to International Computing Technologies (ICT) and £1 million to universities and laboratories.[51] Only a limited share of the funds went

48. Ibid., 155–58; Dosi, "Institutions and Markets," 186.
49. Jeffrey Hart, "West German Industrial Policy," 181.
50. OTA, *International Competitiveness in Electronics*, 152–53. Interestingly, by the late 1980s Nixdorf was in serious trouble and agreed to be purchased by its behemoth competitor, Siemens.
51. Jill Hills, *Information Technology and Industrial Policy*, 155.

to semiconductor research.[52] A Computer Advisory Unit was formed in the government in 1965, and a National Computer Centre was created in 1967.

In 1968 the government encouraged the creation of a single national-champion firm, International Computers Limited (ICL), and all further computer policies centered on keeping ICL afloat. The government created ICL by arranging a marriage between the United Kingdom's only remaining computer manufacturers, ICT and English Electric, with the explicit aim "to establish one strong British-owned computer company able to compete with IBM [and] Honeywell." The Ministry of Technology took a 10.5 percent share of ICL, linked to R&D grants of £13.5 million over four years.[53] By 1969 the government owned 25 percent of the company.[54] Margaret Thatcher's government in 1979 sold off the state's final holding in ICL, which by then was down to 20 percent.[55] Subsidies have been a way of life at ICL. The Tory government contributed £40 million for R&D from 1972 to 1976. Repayment was conditional on pretax profits attaining 7.5 percent of revenues, which never happened.[56]

Problems arose early at ICL because the new company inherited two incompatible mainframe computer lines: the ICT/Ferranti 1900 series and the English Electric System 4, an IBM-compatible machine. The company decided to continue with both lines and develop a new system incompatible with both. The 2900 series (introduced in 1975) had a new operating system designed to facilitate communication between machines (the Virtual Machine Environment), which is still the ICL system. The decision was based more on technology than on markets, and many customers decided to switch to IBM rather than adapt their software to the new series. In fact, ICL's best-selling machines have been the low-end 2903/4 and its follow-on the ME29, essentially successors to the old 1900 series. After 1981 the company's strategy focused on networking, advertising ICL's ability to link its different machines and those of other makers, especially IBM.[57]

Public procurement was also a means of supporting the national computer champion. By 1966 a buy-British policy was in place, as

52. Dosi, "Institutions and Markets," 186.
53. Kelly, *British Computer Industry*, 99.
54. Jeffrey A. Hart, "British Industrial Policy," 153.
55. Kelly, *British Computer Industry*, 74.
56. Ibid., 74.
57. Ibid., 45–46.

Minister of Technology Anthony Benn announced that the government would buy British-made computers "whenever reasonably possible." ICL received 94 percent of central-government orders its first year. IBM's share of central-government orders fell from 32 percent in 1966 to 13 percent in 1968, and its share of public-corporation orders fell from 30 percent to 2 percent over the same years. By 1969 computer purchases by the Ministry of Defense went almost entirely to ICL. The Conservative government of Edward Heath created a Central Computer Agency within the Civil Service Department. For the first time computer-purchasing responsibility for the central government was removed from the spending departments. The buy-British policy was therefore easier to enforce; ICL reportedly was receiving two-thirds of government orders by the mid-1970s. But ICL's shift into the minicomputer market in the 1970s reduced its dependence on government purchases: By 1977 only 7 percent of its business came from government, down from 40 percent (including purchases by public corporations) in the 1960s.[58]

Semiconductors

Europe's difficulties in computers proved to be a drag on the semiconductor sector as well. As Franco Malerba shows in his comprehensive study of the European semiconductor business, the demand for semiconductors in Europe during the 1960s came mostly from the consumer and industrial electronics sectors: Consumer electronics accounted for some 31 percent of final electronics markets in Europe, as compared with only 18 percent in the United States.[59] At first, consumer and industrial semiconductor users were content to replace tubes with discrete transistors. In fact, Siemens in the early 1960s thought that there was "no real demand for integrated circuits."[60]

Later, consumer and industrial users began to replace some discrete devices with linear ICs, not the digital ICs that would become the basic building block of the telematics revolution. Thus, in the 1960s European semiconductor production was led by the traditional, vertically integrated houses like Philips, Siemens, AEG-Te-

58. Hills, *Information Technology,* 156–64.
59. Malerba, *Semiconductor Business,* 122. In the account that follows I rely heavily on Malerba's excellent work.
60. Ibid., 106.

lefunken, Thomson, and General Electric Company (GEC). These large firms produced discrete devices and customized linear ICs for use in their final products, unconvinced that digital ICs would prove broadly useful. Philips and Siemens did not move into production of digital ICs until 1967, and then both relied on licenses from Westinghouse. In general, European companies were late entering the digital IC market and therefore did not benefit either from cumulative learning-by-doing, as the American pioneers did, or from economies of scale, which reward the first producers. By the early 1970s the major European semiconductor makers had retreated from the standard digital IC markets, concentrating instead on small niches not subject to American competition.[61]

Because European computer makers were struggling during the 1960s and 1970s, local demand for digital ICs was limited. By contrast, demand from computer makers became a driving force in semiconductor markets in the United States. Furthermore, because they were struggling, European computer firms were forced to choose between buying second-rate components from European makers or purchasing state-of-the-art circuits from U.S. companies. As using less-advanced ICs would burden them with one more competitive disadvantage, European computer firms bought largely from American IC suppliers. For instance, CII bought chips from advanced American producers rather than from the French semiconductor champion Sescosem, contrary to the *Plan calcul*.[62]

Finally, by the late 1970s European semiconductor firms had begun to try to move back into the standard digital markets. By this time the technology had advanced in miniaturization and integration—into the phases of large-scale integration (LSI) and very large-scale integration (VLSI) (with 1,000–10,000 devices and 10,000–100,000+ devices per chip respectively). Digital ICs were replacing linear circuits even in consumer and industrial applications as well as in the massive telecommunications sector, reflecting the rapid digitization of electronic functions in every branch. Given the increasing complexity in designing and producing advanced circuits, European firms were forced to rely on American and Japanese firms for the technologies. Furthermore, the European governments had begun to support work in digital ICs, having become convinced that

61. Ibid., 119.
62. Ibid., 128, 176–81.

"a domestic productive capability in LSI devices was of strategic value for reasons of national security and for sustaining the electronic and the manufacturing industries as a whole."[63] In other words, governments had decided that microelectronics was a leading sector.

France   Support for the semiconductor industry actually began in France during the first *Plan calcul,* which devoted about one-fifth of its funds to the components industry.[64] In 1968 the government began consolidating pieces of the French semiconductor industry by creating Sescosem within the Thomson group. The new company combined the semiconductor activities of Thomson and CSF, which later merged. The bulk of the *Plan calcul* semiconductor funds went to Sescosem. The government also served as midwife to another new semiconductor firm, Efcis, a joint venture between Thomson and the atomic energy commission.

Two plans in the late 1970s provided focused support for microelectronics. The *Plan informatisation de la société* (1977–80, FFr 400 million) was designed to stimulate the IC industry, the creation of databank services, and the demand for data-processing, especially among small businesses. The *Plan circuits intégrés* (1978–81, FFr 600 million) aimed at enhancing French capabilities in IC design and production.[65]

Further restructuring of the semiconductor industry, again under state tutelage, took place in 1978. The government authorized, negotiated, and helped finance two joint ventures with American firms: Matra teamed with Harris to form Matra-Harris, and St. Gobain linked up with National Semiconductor to form Eurotechnique. The French partner owned 51 percent of the joint venture in each case.[66] The Valéry Giscard d'Estaing government developed a strategy in microelectronics centered on five poles, each specializing in a type of IC (see Table 4.4). The *Plan circuits intégrés* funds were then funneled to these five companies.[67]

Federal Republic of Germany   Siemens began trying to enter the market for LSI chips in the late 1970s, when its management

63. Ibid., 162.
64. Dosi, "Institutions and Markets," 186.
65. Flamm, *Targeting the Computer,* 155; Dirk de Vos, *Governments and Microelectronics: The European Experience,* 45.
66. Dosi, "Institutions and Markets," 191.
67. de Vos, *Governments and Microelectronics,* 44.

TABLE 4.4   FRENCH SEMICONDUCTOR INDUSTRY,
1978

Company	Specialty	Foreign Technology
Efcis	Metal oxide on silicon (MOS)	Motorola
Eurotechnique	Negative MOS (N-MOS)	National Semiconductor (49 percent partner)
Matra-Harris	Complementary MOS (C-MOS)	Harris (49 percent partner)
Radiotechnique	Bipolar for telecoms	
Thomson-CSF	Bipolar for consumer goods and autos	

SOURCE: Organization for Economic Cooperation and Development, *The Semiconductor Industry: Trade Related Issues* (Paris, 1985), 74.

was persuaded that LSI capability was vital for success in the company's core telecommunications and industrial-equipment businesses. At that point the IC division at Siemens had not turned a profit since 1965. The heart of Siemens's strategy was to acquire LSI technology through purchasing advanced American companies. The first major acquisition was a 20 percent stake in Advanced Micro Devices in 1977, followed by 100 percent of Litronix, 100 percent of Microwave Semiconductor in 1979, and 100 percent of Threshold Technology in 1980.[68] In addition Siemens produced Intel microprocessors under license.[69] AEG-Telefunken also manufactured microprocessors, under license from Mostek of the United States, but digital ICs amounted to only about 20 percent of its total IC output.[70]

During this period the federal government funded microelectronics R&D through an electronic-components program. Over the period 1974–78, the BMFT handed out DM 388 million to universities, laboratories, and industry. The program was extended in 1978, and in 1981 it was replaced by a microelectronics program, which funded R&D up to the prototype stage. The OECD cites estimates that semiconductor R&D support by the German gov-

68. OECD, *Semiconductor Industry*, 125.
69. Malerba, *Semiconductor Business*, 156–58.
70. Ibid., 167–68.

ernment amounted to some DM 700–800 million over 1974–83. Again, Siemens received the lion's share of the funding (about 25–30 percent), while AEG-Telefunken took in 10–15 percent, and Valvo (a Philips subsidiary), about 10 percent.[71]

United Kingdom   Support for the semiconductor industry in the United Kingdom began in a limited way during the late 1960s, with a small share of the Advanced Computer Technology Project's funds going to IC pilot production. Focused aid to the industry began in 1978 with two new programs: the Microprocessor Applications Project (MAP) and the Microelectronics Industry Support Plan (MISP). MAP aimed at speeding the diffusion of microprocessor applications throughout industry by funding training projects, workshops for company executives and workers, and programs to introduce computers into schools. The MAP budget totaled £55 million for three years (later extended). MISP focused on commercial production of both standard and application-specific ICs, with grants of 25 percent of approved project costs (raised to 33 percent in 1982) and cost-sharing contracts with 50 percent assistance. The program also placed preproduction orders for IC manufacturing equipment, thus accelerating the acquisition of advanced production machinery. The MISP budget for its first five years was £55 million, and the program was extended for 1984–90 at £120 million. The U.K. semiconductor industry also received support from the Electronic Component Industry Scheme, which distributed £20 million over 1977–80.[72]

In addition, Britain took the bold step in 1978 of creating a semiconductor start-up company to produce standard VLSI memory and microprocessor chips. At that time no British company manufactured standard ICs. The four main British electronics firms (GEC, Plessey, Ferranti, and Standard Telephone and Cable (STC)) produced custom and semicustom circuits for niche markets, mainly in telecommunications and defense. The National Enterprise Board agreed to finance the launching of Inmos, with a commitment of £50 million in government money. The Thatcher government, initially uncertain as to the stance it should take regarding Inmos, re-

71. OECD, *Semiconductor Industry*, 72–73; de Vos, *Governments and Microelectronics*, 76–77.
72. OECD, *Semiconductor Industry*, 71–72, 80; de Vos, *Governments and Microelectronics*, 25–26.

leased the second installment of the approved funds in July 1979. By 1984 Inmos had received from the government a total of £65 million in cash and £35 million in loan guarantees. That year also saw the first profits for Inmos, which then was sold to Thorn-EMI as part of the Thatcher privatization program.[73]

Telecommunications

If microelectronics and computer strategies in Europe have been vigorously nationalistic, telecommunications policies have been even more so. Until the late 1980s the European telecommunications agencies, or PTTs,[74] kept a tight grip on managing the network, providing services, and regulating (sometimes even providing, installing, and maintaining) customer-premises equipment (including phone handsets, modems, private branch exchanges, and facsimile machines). This position of power gave the PTTs a dominant role vis-à-vis the manufacturers of equipment of all kinds: makers of transmission equipment and cable, public switches, and terminals all marched to the cadence set by the PTT. Indeed, the PTT was an important instrument of industrial policy.

In this section I do not attempt to catalogue national telecoms policies, descriptions of which could easily fill a chapter or two apiece. Rather, I summarize and illustrate with examples the major elements common to most European telecommunications systems in the early and mid-1980s. I will also highlight the most significant differences among states.

Networks   The telecommunications network is the infrastructure for communications, linking on demand customers in diverse places, demanding diverse services. In a sense, the network infrastructure is like a country's system of roads and highways, which connect every location to every other and permit different kinds of usage (trucks, cars, buses, bicycles). In Europe governments have regarded the network as a public utility, and network provision and management as the domain of public monopolies. The United States

73. Malerba, *Semiconductor Business,* 170; OECD, *Semiconductor Industry,* 72; Hills, *Information Technology,* 199–217; W. B. Willott, "The NEB Involvement in Electronics and Information Technology," 210–12.
74. Presently in Europe the acronym NTO is in vogue, short for National Telecommunications Organization, which reflects the new role of telecoms—more than and different from mail, telegraph, and phones. In this study I employ the old label.

has numerous overlapping, interconnected, public and private telecoms networks.

In Europe, as a rule, the PTT has owned the network (switching
and transmission equipment), and all services had to be offered via
that network. For instance, in Germany the Bundespost possessed
by legislative fiat the "exclusive right to build and maintain telecommunications facilities."[75] The Bundespost could authorize private networks as long as these were not connected to the public
network and were used only for internal communications. Thus the
data-transmission and teletex networks in Germany were owned
and operated by the Bundespost.[76] In France the state also maintained a monopoly on the network. In fact, the Direction Générale
des Télécommunications (DGT) until the early 1980s played a highly
autonomous role in directing the modernization of the French telecoms network. France did not permit the shared use or resale of
lines leased from the DGT, maintaining the public network monopoly and foreclosing one of the avenues used by American companies to offer new services and private networks.[77]

Great Britain offered a contrast to the European norm, though
its deregulation of the network has been far more cautious and limited than that seen in the United States. In 1986 the government
authorized a private company, Mercury, to provide telecommunications services on its own network, ending the network monopoly
begun in 1912. However, no more competitors could be licensed
before November 1990.

Equipment   Logically, because the European PTTs have been
sole caretakers of the network (except in Great Britain), they also
constitute a monopsony for telecommunications equipment. Manufacturers of transmission equipment and switching equipment have
historically been intimately linked, therefore, with their national
PTTs.[78] Even the markets for terminal equipment have been subject

75. Patrick Cogez, "Telecommunications in West Germany," 46.
76. Ibid., 49.
77. CEC, *Towards a Dynamic European Economy: Green Paper on the Development of the Common Market for Telecommunications Services and Equipment*, 83.
78. Transmission equipment is that which carries the signal between terminals
and switches, and includes coaxial and fiber-optic cable, microwave radio, and satellites. Switching equipment makes the electronic connections between terminals
and comprises exchanges in central offices. I will call the two together network
equipment.

to PTT control.[79] For instance, the United Kingdom's 1969 Post Office Act gave the British Post Office (BPO) the exclusive right to produce, install, and maintain network equipment. The BPO even had the power to specify which private automated branch exchanges (PABXs)[80] and modems could be attached to the network, with a monopoly on installation and maintenance.[81]

Similarly, according to the law chartering the Bundespost, terminal equipment was considered part of the network. Therefore, manufacturers could sell directly to customers only on authorization by the Bundespost. By the early 1980s the Bundespost still retained its monopoly on provision of the first telephone handset (the administration bought the handsets from private makers and resold them) and on all modems.[82] In France the market for terminal equipment has been open in principle since 1920, though all equipment had to be approved for connection by the PTT. In practice, the customer-premises-equipment (CPE) market was quite open, with a few exceptions. Videotex terminals were supplied solely by the PTT. However, the PTT sold only small PABXs (twenty lines); for anything larger, customers could choose among private suppliers.[83] Naturally, the requirement of PTT approval was used in each country to favor national manufacturers. It is also worth noting that in the European countries that did not have strong indigenous suppliers of telecoms equipment,[84] the markets tended to be supplied by multinational corporations, notably ITT, Ericsson (Sweden), and Siemens.

Concerning network equipment, PTT control of the network provided the means to carry out industrial policies regarding the telecoms-manufacturing sector. As with semiconductors and computers state-arranged marriages of telecoms companies have not been unusual. For instance, the British Industrial Restructuring Council

79. Terminal equipment is that "into which the original signal is introduced and from which a final signal can be received" and is generally placed on the customer's premises. Terminals include telephone handsets, telex and facsimile machines, and teletex terminals. From OECD, *Telecommunications*, 19.

80. PABXs are the switchboards used in a large building or company to route calls to and from its many phones and terminals.

81. Hills, *Information Technology*, 118–19.

82. Cogez, "Telecommunications in West Germany," 52.

83. Ibrahim Warde, "French Telecommunications," 99–100.

84. Austria, Belgium, Denmark, Finland, Ireland, Norway, Spain. See Michael G. Borrus et al., *Telecommunications Development in Comparative Perspective: Appendix*, 41.

pressured GEC to take over AEI.[85] Also, before the 1980s switching-equipment purchases by the BPO were allocated among members of the "ring" of suppliers. The original deal was struck in 1924 with five suppliers. By 1969 the number was down to three (GEC, Plessey, and STC, an ITT subsidiary), and their shares of orders were fixed at 40 percent, 40 percent, and 20 percent, respectively. In 1982 STC was dropped as a supplier of the digital System X exchange.[86]

A similar process of concentration took place in France. In the mid-1970s three companies supplied the French switching market: CIT-Alcatel (a subsidiary of Compagnie Générale d'Electricité (CGE)), Compagnie Générale de Constructions Téléphoniques (CGCT) (a subsidiary of ITT), and Thomson Telecommunications (formed by the DGT by merging an Ericsson subsidiary with an ITT subsidiary).[87] All three were nationalized by Mitterrand in 1982. The following year, the government merged the Thomson and CGE telecoms activities into one company, Alcatel, within the CGE group. The DGT opposed this move because it had been pursuing a competitive bidding policy with competition between CGE and Thomson. Alcatel at this point held about 80 percent of the French switch market. The champion grew further in 1986 with the purchase of ITT's European telecommunications businesses.[88]

The German telecommunications-equipment industry has not passed through the concentration process visible in France and the United Kingdom; Siemens was already the national champion. Siemens took a 46 percent share of the domestic central-exchange market in 1978, compared with 30 percent for Standard Elektrik Lorenz (SEL) (an ITT subsidiary, later transferred to CGE), 14 percent for DeTeWe, and 10 percent for Telefonbau und Normalzeit.[89] Siemens has traditionally worked closely with the Bundespost in developing equipment and setting standards.[90]

As a consequence of PTT equipment-approval requirements and the intimate liaisons of the PTTs with national producers, European

85. Rob van Tulder and Gerd Junne, *European Multinationals in the Telecommunications Industry*, 46.
86. Ibid., 45–46.
87. OECD, *Telecommunications*, 54–55; Jeffrey A. Hart, "The Politics of Global Competition in the Telecommunications Industry," 182.
88. van Tulder and Junne, *European Multinationals*, 49–50; Hart, "The Politics of Global Competition," 182.
89. Cogez, "Telecommunications in West Germany," 70.
90. See, for example, Hart, "The Politics of Global Competition," 187.

telecoms-equipment markets have until recent years been essentially closed to nonnational producers. This restriction applied to non-European firms, like IBM and Northern Telecom (Canada), as well as to powerful European companies that have Europe-wide and worldwide operations, like Ericsson and Philips. For instance, the Bundespost invited Philips to compete in digital exchanges only in 1981. IBM got its first major break in Germany in 1984, when it won the contract to provide the Bundespost with equipment for its videotex system, *Bildschirmtext,* in competition with Siemens.[91]

In France the major subsidiaries of foreign companies (belonging to Ericsson and ITT) were both nationalized by 1982, though CGCT (formerly of ITT) was later sold to Ericsson, giving the Swedish firm CGCT's 16 percent share of the French digital-exchange market.[92]

Foreign-based telecoms gained access to the U.K. public switch market in the early 1980s but in only a limited way. British Telecom (BT) signed an agreement with Mitel for digital-exchange technology in 1981 and bought a single public exchange from IBM in 1982.[93] BT also announced it would accept tenders for public exchanges from Northern Telecom, Thorn-Ericsson (joint venture), and Pye-TMC (a Philips-AT&T joint venture), with the intention of buying 10 to 20 percent of its digital exchanges from non-British suppliers.[94] Foreign firms have made the most headway in PABXs. The BPO permitted Ericsson to start selling its PABX in 1969; by 1974 Ericsson had a 20 percent market share. IBM followed, gaining by 1983 a 16 percent market share in PABXs with over 100 lines.[95] The key in PABXs was that Ericsson and IBM developed the digital technology in advance of U.K. firms. GEC markets PABXs under license with Northern Telecom, and Plessey under license with Rolm.[96]

The upshot is that each European market for network equipment has traditionally been supplied by an oligopoly dominated by a single domestic maker working closely with the PTT. Import penetration ratios for telecommunications equipment in 1981 were low—

91. van Tulder and Junne, *European Multinationals,* 55.
92. Robert Gallagher, "Suddenly, the Rules Change for Europe's Telecom Business," 114.
93. Hills, *Information Technology,* 85.
94. François Bar, "Telecommunications in the United Kingdom," 82.
95. Ibid., 87.
96. Hills, *Information Technology,* 85, 122.

2.0 percent in France, 4.1 percent in Germany, and 10.4 percent in Britain.[97] In addition each PTT selected a different network standard, meaning that switches made for one national market must be converted at significant cost to be sold in another. Even the terminal-equipment markets were slanted in favor of domestic suppliers by PTT approval requirements.

Services   Given the far-reaching sovereignty of the PTTs over telecommunications networks and equipment, it should not be surprising that the PTTs should also exercise dominion over services: what services can be provided and by whom. In Britain the old BPO held a monopoly on provision of services, although a 1982 law opened up the market for provision of value-added networks (VANs).[98] France's DGT took the initiative in introducing new services. Some liberalization began in the early 1980s. The operating principle of the DGT was that it would retain the network monopoly as well as telex and facsimile systems, but the VAN market would be relatively open (foreign companies can offer services only via joint ventures with French firms).[99]

The legal mandate of the Bundespost included telecommunications services, and it was extremely cautious in approving new service providers. For instance, whereas businesses in the United Kingdom had access in 1985 to over 400 VANs, those in Germany could tap into only about a dozen.[100] German law also required that a certain minimum percentage of any data-processing that involved telecommunications transmission be carried out in Germany. This law restricted access by German users to non-German databases and some kinds of advanced VANs.[101]

97. van Tulder and Junne, *European Multinationals*, 13, 40.

98. VANs are the means by which companies provide enhanced services—that is, services beyond basic voice and data transmission. Value-added services literally "add value to data communications by providing file storage, message switching, protocol conversion [translation between machines operating on different standards], ... and access to database and other information services." OTA, *International Competition in Services*, 159. Videotex services are VANs that permit access to various kinds of information services, like stock quotations and airline schedules.

99. Ibid., 173.

100. Joseph Fitchett, "Europe Sits by the Phone, Awaiting a Revolution," *International Herald Tribune*, 4 December 1985, p. 7.

101. Hart, "The Politics of Global Competition," 187.

TABLE 4.5.   PTT MONOPOLIES, 1984

	Networks	Services	Terminals	Standards	Tariffs
Belgium	M	M	M	M	M
Denmark	M	M	M	M	M
France	M	M	S	M	M
Germany	M	M	M	M	M
Greece	M	M	M	M	M
Ireland	M	M	S	M	M
Italy	M	M	M	M	M
Luxembourg	M	M	S	M	M
Netherlands	M	M	M	M	M
United Kingdom	S	S	O	S	M
United States	O	O	O	O	O

M = monopoly
S = semi-open
O = open

In short, European PTTs exercised broad monopoly powers in the early 1980s. Table 4.5 depicts the situation as of 1984.

SUMMARY

In the immediate postwar period national governments in Western Europe assumed a central role in orchestrating the rebuilding of industrial economies. That role included channeling capital into sectors deemed crucial as a foundation for economic growth, usually basic industries like steel, cement, and energy. State involvement in promoting economic development and shaping the industrial composition of the national economy became fixed in the welfare democracies that took shape in Europe over the succeeding decades.

I have shown how science policy became wedded to economic policy in the early 1960s, as governments began to expect returns to the economy at large from public investments in scientific research. The result was a series of national technology ministries and technology policies. Beginning in the late 1960s technology policy began to be linked to industrial policy. This linkage stemmed from the growing recognition of the role of high-technology industries in driving economic growth. Fears inspired by the technology-gap scare moved governments to establish national-champion computer companies. The drive for national champions in semiconductors began in the 1970s. Telecommunications was under the close supervision of the PTTs throughout the period. Telematics policies were an instrument of national governments to sustain national employment and economic growth, on which electoral success depended. It was only natural, therefore, that the first telematics policies should bear a unilateralist stamp.

FIVE

Successes and Failures

Collaboration in the 1970s

Europe experimented with technological collaboration during the 1970s. A brief look at the successes and failures of that decade will provide a necessary historical foundation for analysis of telematics collaboration in the 1980s. In this chapter I dissect two instances of failure to cooperate in telematics, one a Commission (CEC) initiative and the other the Unidata consortium. I also look briefly at two successful collaborative ventures in Western Europe during the 1970s, Airbus and the ESA. Comparisons between the aerospace successes and the telematics failures will permit some revealing insights into the dynamics of international technological collaboration.

GETTING NOWHERE: TELEMATICS COOPERATION

Given the commitment of European governments to national economic growth and the resulting nationalistic telematics policies, it is not surprising that calls in the 1970s to rally Europe's diverse countries behind a collaborative banner seemed to fall on deaf ears. National-champion strategies were in place.

EC Efforts

Until the second half of the 1970s telematics projects at the European level were subsumed under the Commission's attempts to establish comprehensive science and technology policies under its own auspices. The OECD's technology-gap reports gave the first nudge to science and technology policy initiatives at the European level.[1] The Comité de Politique Economique à Moyen Terme (CPEMT), created by the EC at the request of the French in the spring of 1964, had as part of its brief a charge to study the potential for a Common Market science and technology policy. CPEMT created a subcommittee, PREST (Politique de Recherche Scientifique et Technologique), in March 1965 for this purpose. In addition, the three ECs formed a joint committee to study Community-level R&D.[2] Also at about this time British Prime Minister Wilson proposed a European Technological Community. This proposal coincided with intense interest in advanced technologies within the Labour government but was also a way of highlighting a real strength the British could bring to the Common Market; at the time British technological capabilities were thought to be at the forefront in Europe. In any case, de Gaulle vetoed British entry, and the idea slipped away.[3]

The PREST report was delivered at the first meeting of science ministers (October 1967). PREST was subsequently requested to look into possibilities for cooperation in seven specific sectors: information sciences (computing), telecommunications, transport, metallurgy, meteorology, oceanography, and environmental protection. The Council of Ministers instructed PREST to consult with private industry in formulating project proposals. During the following year French intransigence in denying participation in PREST meetings to four countries then applying for EC membership led to a cessation of work in the committee. Eventually the French and Dutch foreign ministers decided on an acceptable compromise, and the committee resumed its deliberations under a new chairman, Pierre Aigrain. The Aigrain Report was submitted to the CPEMT, the

1. I rely for this summary of EEC R&D programs during the 1960s and 1970s on three excellent accounts: N. H. Aked and P. J. Gummett, "Science and Technology in the European Communities: The History of the COST Projects"; Haas, *The Obsolescence;* and Henry R. Nau, "Collective Responses to R&D Problems in Western Europe: 1955–1958 and 1968–1973."

2. Aked and Gummett, "Science and Technology," 273–74.

3. Other proposals emerged for European R&D collaboration during that period. Amintore Fanfani of Italy suggested a "technological Marshall Plan." Christopher Layton proposed technological cooperation at the European level, as did another offering from the United Kingdom, the Plowden-Winnacker Plan. See Roger Williams, *European Technology: The Politics of Collaboration,* chap. 2.

Council, and the Comité des Representatives Permanents (CORE-PER). This report advanced forty-seven specific research proposals under the seven headings. The Aigrain Report was passed on favorably by the COREPER and then by the Council in October 1969. In addition, the Council decided to send the report (now containing thirty project proposals, the other seventeen having been dropped for lack of interest) to nine countries not members of the Common Market.[4] All nine indicated an interest in participating. Still, no action was taken beyond meetings of experts from the fifteen countries in 1970, even though the Hague summit had produced a statement in favor of increased common R&D programs. Eventually a number of PREST projects worth $30 million were launched but without non-EEC partners.[5]

In October 1970 the Council created the COST committee, composed of senior science and technology officials from the fifteen countries. The Council itself provided a secretariat, but the Commission was not represented on the committee. Science ministers from nineteen countries (Finland, Greece, Yugoslavia, and Turkey had joined) met in November 1971 and agreed on seven projects worth a total of $21.5 million (from the thirty proposed in the Aigrain Report). Each project entailed a separate agreement among those states desiring to participate. COST projects involve no common funds. States decide which projects they want to join, and the contribution of each participant is spent entirely in its own laboratories and industries. Research undertaken in corporations must be underlying, or far from the market (precompetitive in today's jargon). Out of twelve COST projects listed by N. H. Aked and P. J. Gummett (up through mid-1974) four dealt with telematics (data-processing and telecommunications).[6] The point is that what began as an effort to devise EC science and technology policies ended up as a small, minimally funded collection of intergovernmental research agreements.

With a new commissioner for industry and general research and technology, Altiero Spinelli, the Commission made ambitious new proposals for joint R&D. It asked the Council to recognize EC competence in all areas of R&D (not just those named specifically

4. The United Kingdom, Ireland, Denmark, Norway, Sweden, Austria, Switzerland, Spain, and Portugal.
5. Nau, "Collective Responses," 631.
6. Aked and Gummett, "Science and Technology," 289–90.

in the Treaty of Rome—namely, coal and steel and atomic energy). In plans released in 1970 and 1972 Spinelli also proposed a European Research and Development Committee, a European Science Foundation, and a European Research and Development Agency. The European Research and Development Agency (ERDA) would have its own budget to carry out R&D projects, and the Commission itself would have an R&D budget of 120 million units of account (UA).[7]

None of these plans got very far.[8] As Ernst Haas put it, the proposals from the Spinelli era "failed essentially because of the Commission's insistence on adding the R&D question to the package of steps which was to lead toward political union."[9] Spinelli's vision included a dramatic strengthening of EEC institutions. "Mainly because of the centralisation implied by these proposals . . . they were not well received in national capitals," according to Aked and Gummett.[10] In other words, at this stage, EC-level actions on behalf of telematics were embedded within initiatives for a broad EC mandate in science and technology, which were in turn linked to dramatic upgrading of EC institutions.

A new commissioner for science, research, and education, Ralf Dahrendorf, shifted the Commission approach away from ambitious Community-level institutions and R&D policies. In the summer of 1973 Dahrendorf proposed that the EC compare and coordinate national R&D policies through a committee of high-level government representatives called CREST (Comité de la Recherche Scientifique et Technique). The Council approved in January 1974. CREST was established as a joint committee of the Council and the Commission and took over the duties of PREST. The most significant point is that CREST abandoned the ambition for a common R&D policy in favor of coordination of national policies. As in COST member states participated in only those projects that interested them (the variable-geometry, or à la carte, approach). Furthermore, the work of CREST was linked to broad policy areas like

7. The unit of account was an EEC creation defined on the basis of a basket of currencies; one UA is now equal to one ECU.
8. The European Science Foundation, which came into being later, was not an EC body but rather an association of academic and research councils from its member countries.
9. Haas, *The Obsolescence*, 55.
10. Aked and Gummett, "Science and Technology," 282. Nau concurs; "Collective Responses," 638.

the environment, energy, and industry, as opposed to specifically economic objectives. One CREST subcommittee dealt with data-processing. Far-reaching science and technology policy coordination has never emerged in CREST, though CREST continues to exist as an advisory body to the Commission and the Council. In sum, from Spinelli to Dahrendorf, Commission ambitions were scaled down, and telematics remained a small, insignificant part of the total package.

Unidata

At this point the ill-fated Unidata venture enters the picture. Unidata was a consortium involving Siemens (West Germany), Philips (the Netherlands), and CII (France) in an attempt to combat IBM's dominance of European and world computer markets.[11] The story begins in 1971, when RCA withdrew from the computer business. Siemens, which had been relying on RCA licenses for mainframe technology, suddenly found itself in need of a new source of technology. CII suggested it could supply that need. Discussions began. In early 1972, when it appeared that Siemens and CII were close to a partnership agreement, Philips declared a strong interest in joining the project. After thorough negotiations, the three firms announced in July 1973 that they would collaborate to develop and market a full range of computers, which would carry the Unidata label.

The consortium aimed at providing a family of mainframe computers (that is, a set of machines ranging from small to large that could run the same software). This plan was seen as crucial for competing with IBM, which had pioneered the family approach. (Still, the three companies together accounted for only 9 percent of the European computer market.) Within Unidata, Philips (which had abandoned large computers) would supply the two smallest mainframes, Siemens and CII would share the mid-range machines, and CII would produce the largest two mainframe models. Siemens and Philips were building and selling computers by 1975, but CII was lagging on its development and production plans. The biggest problem was organizational. So far, the work of the consortium had

11. For this account of Unidata I rely on two unpublished manuscripts by Joseph M. Grieco: "Technical Knowledge, Rational Exchange, and International Cooperation: The Cases of Airbus Industrie and the Unidata Computer Consortium" and "Toward a Realist Understanding of International Cooperation."

been conducted through a number of interfirm working committees. This system proved slow and uncoordinated, and the three partners decided they needed a joint headquarters.

At this point CII sought government support for the new Unidata structure to be set up in Brussels (remember, CII was created and partly owned by the French government). The government was in the midst of a debate over the future of the French computer industry. One option being considered was a merger of CII with Honeywell-Bull, thereby recovering Bull from its American owner. The other main option was to have CII continue with Unidata, with the possibility of CII's eventually purchasing Honeywell-Bull itself. The government was divided during 1973 and 1974, though favoring the second alternative. But with the new government of Giscard d'Estaing, the merger camp prevailed. In May 1975 Paris announced the merger of CII and Honeywell-Bull.

In short order Siemens and Philips declared that CII-HB was no longer welcome in Unidata. CII-HB would be a trojan horse, carrying an American company (Honeywell) straight into the center of Europe's supposed champion. After all, Unidata was to be the European answer to American domination of the computer industry. Though the Dutch government wanted the remaining two companies to continue together, the German government was ambivalent. Eventually Philips announced its intention to withdraw, and by December 1975 Unidata was officially disbanded. The experience left a legacy of considerable bitterness and skepticism concerning European cooperation in information technologies. In fact Unidata came up repeatedly during discussions of ESPRIT in the early 1980s.

The Commission Tries Again

After the collapse of Unidata, the Commission had to regroup. It had expressed its high hopes for the European consortium and now began to seek other means for building up the telematics industries in the Community. The Commission sought about 8 MECU in late 1975 for computer projects, including one-third of the total for an advanced real-time computer language. This was an area in which Europe was thought to have a lead over the United States. But by August of 1977 the proposed budget had shrunk to 3.6 MECU as a result of German pressures, and the Council still had not reached a decision. The major problems were that France wanted close su pervision of the EC project by national governments, and there was no agreement on the nationality of the project leader. The Germans were said to be particularly intransigent on this issue. Still, the Commission was able to gain approval for an ad hoc handful of computer projects.[12]

The Commission's first programmatic initiative came in 1976, when it proposed a program to support R&D in the data-processing sector. Member governments were unenthusiastic, and the Council dallied. The Council did not approve the Multiannual Data-Processing Program until July 1979. Even then it reduced the Commission's proposed budget from 108 MECU to 25 MECU, to be spread over four years. The program included research on standardization, finding common procedures for public procurement, studies of technological needs and possible EC programs, and studies of the effects on employment, cooperative R&D, and market opening. Collaboration among firms from different states was to be encouraged by requiring that projects involve more than one state. Hardware work was to focus on minicomputers, microcomputers, and peripherals; a major share of the software effort would focus on a common programming language. Any software developed in the program was to be portable.[13] The plan provoked objections from European computer makers who said they should have had more voice than they did in its preparation and who complained that by the time the Council got around to the program it would be "too little, too late."[14]

In short, Commission efforts during the 1970s to mobilize technological collaboration did not spark bursts of enthusiasm. Even though the technology-gap debate of the 1960s pointed to a crisis in Europe's telematics industries, collaboration to support development of the threatened sectors did not emerge. Isolated telematics projects were buried in proposals for EC science and technology planning or policy coordination, which member states rejected. The R&D programs that did obtain approval (COST, CREST) did not involve joint industrial strategy but rather ad hoc, intergovernmental research agreements. In the latter half of the 1970s the Commission proposed a program for information technologies that the

12. "Babel Still Rules," *Economist*, 6 August 1977, p. 44.
13. *Agence Europe*, 25 July 1979, p. 6; *Agence Europe*, 1 April 1983, p. 12.
14. "Buy European," *Economist*, 21 May 1977, p. 59.

Council finally approved—after a delay of three years and at a severely reduced level of funding.

As the previous chapter showed, national-champion strategies flourished in the major European states during the 1970s. European governments would consider technological collaboration only after national policies had failed, thereby triggering a process of policy adaptation.

SUCCESSES IN AEROSPACE

Civil aerospace provides two examples of European technological collaboration that took root after national approaches had not panned out. In the cases of passenger aircraft and rocket launchers, national leaders were in an adaptive mode, willing to consider something besides national strategies. In both sectors European industry instigated or at least strongly favored cooperation. Both cases demonstrate the importance of organizational arrangements that permit flexibility in participation and favor a satisfactory distribution of costs and benefits. In the following pages I leave telematics aside for an excursus into two aerospace collaborations.

Airbus

Civilian aircraft constitute an important high-technology sector, both as producer and major consumer of innovations in a variety of fields: new materials, aeronautics, electronic communications and navigation systems, engines. Table 3.1 showed aerospace to be the most research-intensive of all industries. In addition, aircraft companies are major employers and produce for a civil market that is large and growing: In 1985, 654 aircraft worth about $23 billion were ordered.[15] McDonnell Douglas (MDD) (the second largest U.S. producer of civil aircraft, behind Boeing) estimated in 1987 that over the next decade and a half, 5,400 aircraft would sell for some $250 billion.[16] Finally, governments are interested in civil aircraft to maintain the skills needed for military planes.

The market for civil aircraft has been dominated by the United States. American companies by 1985 had supplied 83.7 percent of

15. Keith Hayward, *International Collaboration in Civil Aerospace*, 22.
16. Cited in Thomas A. Sancton, "Airbus Takes Wing," 40.

all civil jet aircraft delivered in the West.[17] Boeing alone held 60 percent of the world market in that year.[18] The primary struggle for the European aircraft industry has been to establish itself in the face of Boeing's dominance, much as European computer makers have had to contend with the leviathan IBM. Cooperation began after national strategies for meeting the challenge proved unviable.[19]

The Germans, excluded from the aircraft industry in the aftermath of World War II, saw Airbus as the only way to bring their industry technologically up to date. Tying their aircraft rebirth to European ventures was politically expedient. The British and the French had been unable to establish viable civil-aircraft industries in the emerging jet age. Britain's Comet proved a disastrous failure (the plane would not stay in the air because of structural defects), and France's Caravelle (with Rolls-Royce engines) never got anywhere on the market. Thus, the Plowden Report in Great Britain concluded in 1965 that the only hope for British civil aviation lay with European collaboration.[20] The French reached a similar conclusion, and Airbus was born. Interestingly, Airbus began to take shape just as disenchantment with the Concorde was reaching tense new highs. But aircraft policy-makers learned some lessons from the Concorde. Airbus would be run by industry, not by intergovernmental committees. Its target would therefore be the market, not technological and political prestige.

Airbus began with discussions among governments (France, Britain, and Germany), aircraft builders, and airlines in 1966. The outcome of the meetings was the formation of an industrial consortium, with one company from each country, to design a European jetliner.[21] Early design work, which required shepherding the project through intense disagreements over the plane's specifications (size and engine selection), was led by two executives of Sud Aviation (later to become Aerospatiale). French leadership, as Britain withdrew from the project, proved decisive.[22] Sud Aviation and Deutsche

17. Hayward, *International Collaboration,* 23.
18. Ibid., 27.
19. John Newhouse, *The Sporty Game,* 123–24; Hayward, *International Collaboration,* 36–37.
20. Newhouse, *Sporty Game,* 123–26.
21. Ibid., 125.
22. Ibid., 190–91.

Airbus—a German group with Messerschmitt-Bölkow-Blohm (MBB) as the major partner—were the founding industrial partners when Airbus was established as a *groupement d'intérêt économique* (GIE) under French law in December 1970. The British firm Hawker-Siddeley was retained to produce the wing, with the help of a subsidy from the Germans for the design work.

The initial product for Airbus was the A-300, a twin-engine, wide-bodied plane for medium to long hauls, with a capacity of 250–70 passengers, a market niche none of the Americans was producing for. In the mid-1970s Airbus planners decided that in order to survive they had to produce a product family, as Boeing did (and as IBM did in computers). A family of aircraft offers to airline companies economies in parts, maintenance, and training. The result was the A-310, a twin-engine, wide-bodied, short- to medium-range plane for 200 passengers. The British officially rejoined the consortium in 1979 for the A-310. British Aerospace became a risk-sharing partner in Airbus with a 20 percent stake. Since then, the home governments of Airbus members have approved the loans and guarantees needed to add new members to the family, the short-range, 150-seat A-320; the medium-haul A-330; and the long-haul A-340.

With its growing range of offerings Airbus has emerged as a significant competitor to Boeing and MDD. By 1984 Airbus had delivered 252 A-300s and 46 A-310s (which entered the market in 1983). Airbus accounted for 18 percent of all deliveries in 1984, while Boeing delivered 54 percent of all aircraft and MDD, 17 percent.[23] In its niches, though, Airbus was very strong: nearly 60 percent of the 1984 market for twin-engine wide-bodies.[24] The A-320 was a bold advance. While the earlier two models had been built with existing technologies, the A-320 embodied several state-of-the-art advances. It was the first commercial jetliner to have fully electronic controls for the flaps and ailerons (as opposed to wires and hydraulics). The cockpit, with video-display panels, required a crew of two as opposed to the standard three.[25] Boeing and MDD did not yet have new direct competitors to the A-320. By mid-1987 the

23. Data from "The Survivors," *Economist,* 1 June 1985, Survey, p. 8.
24. Ibid., p. 9.
25. See Jacques Morisset, "L'Airbus A.320, héraut de l'industrie européenne," 69–70; Sancton, "Airbus," 43, 46.

A-320 had tallied over 440 orders and options, making it the "fastest selling plane in aviation history."[26] The two newest planes, the A-330 and A-340, had attracted more orders and options by late 1990 than their most direct competitor, MDD's MD-11, by 400 to 379, with a lead in firm orders of 210 to 158.[27]

In its organization Airbus Industrie is an odd beast. As a GIE, it is neither a joint company nor a merger. All technological and production work is in the hands of member companies, which since 1970 have been Aerospatiale (37.9 percent share), Deutsche Airbus (37.9 percent), British Aerospace (20 percent), Construcciones Aeronauticas S.A. (CASA) of Spain (4.2 percent). These companies are the owners (with voting power proportional to national shares) as well as the principal subcontractors for Airbus. Airbus Industrie coordinates work among the contractors and handles sales and product support. So far the distribution of contracts has strayed somewhat from the percentages of national shares, favoring the French.[28] Increasingly, however, national shares are hard to specify because many of the components are supplied by multinational joint ventures and Airbus encourages bids by consortia.[29] American parts figure substantially in Airbus planes: up to 40 percent of the costs for an A-300 or A-310 (largely because of reliance on U.S. engines), but declining to about 8 percent of the A-320. Subassemblies are manufactured at the contracting plants all over Europe, with final assembly (representing only 4 percent of the total work) at Airbus headquarters in Toulouse, under the direction of Aerospatiale.[30]

The role of the governments is limited to oversight by two committees (the Intergovernmental Committee and the Airbus Executive Committee) composed of representatives of the relevant ministries. An Airbus Executive Agency acts as the liaison between Airbus and the governments on a continuous, working basis. The studies behind each new model are carried out by the companies themselves, which then work with the Airbus Executive Agency to devise a framework agreement. A framework agreement, once approved

26. Sancton, "Airbus," 40.
27. Ibid., 45; "Deliveries Begin of Fuselage Sections for A330/A340 Aircraft," 29; Carole A. Shifrin, "Airbus Industrie Expects To Make Profit This Year for First Time," 67.
28. For instance, on the A-310, French companies received 47 percent of equipment contracts by value, while British firms received 11 percent and German firms 30 percent. See Hayward, International Cooperation, 91.
29. Ibid., 77.
30. Ibid., 71.

by the Intergovernmental Committee, specifies the national shares in each Airbus model. Still, each industrial partner must negotiate with its government for the loans and loan guarantees needed to launch a new aircraft.[31]

The murkiest area of Airbus is that of finances and accounts. Because of its legal status as a GIE, Airbus does not have to publish financial reports. Therefore, outsiders (and perhaps insiders) have no way of knowing what Airbus's costs are and therefore what its profits are. Estimates of total government funding for the first three models (the A-300, A-310, and A-320) range from $7.2 billion[32] to $15 billion.[33] The two new planes were launched with an additional $4 billion in government financing.[34] Still, as Airbus proponents are quick to point out, American aircraft makers have long benefited from defense R&D contracts, Pentagon purchases, and tax breaks. Furthermore, Airbus decided in early 1989 to reorganize its finances and to open, eventually, its accounts.[35]

What lessons derive from the Airbus experience? First, industry has played a central role both in instigating cooperation and in carrying out the technological and industrial planning. In a collaborative venture that involves commercially relevant technologies, industry must have a central role in defining the project. The Concorde, for which intergovernmental committees managed the work, provides a counterpoint to Airbus in this regard. Second, the organizational arrangement has allowed the various partners to participate at the level they desire. With each new Airbus model, the partners commit to the share they feel will make the venture worthwhile. But these shares are then fixed for the life of that model, which could run into decades. Different shares are possible with new aircraft.

ESA

The exploration and exploitation of space require the coordinated application of dozens of advanced technologies. Space R&D thus

31. Ibid., 69–70.
32. Ibid., 52.
33. Sancton, "Airbus," 44.
34. Axel Krause, "Airbus President Brooks No Doubt," *International Herald Tribune,* 18–19 April 1987, Finance Section, p. 1.
35. See, for example, Editorial, "Accounting for Airbus," *Financial Times,* 11 March 1986, p. 24; "Airbus Moves for Efficiency," *Wall Street Journal,* 22 March 1989, p. D22.

produces as well as consumes innovations in microelectronics, computers, advanced materials, telecommunications, exotic chemicals, radar, and other technologically advanced sectors. Indeed, the space sector is intimately linked to major developments in other sectors, most obviously telecommunications. The advent of satellites helped transform communications into the dynamic, leading technology sector that it is today.

The exploitation of space is also a substantial economic sector in its own right, though at nowhere near the level of computers or civil aircraft. The OECD notes that between 1970 and 1980 thirty-seven telecommunications satellites were launched in the West. For the period 1985–90 an average of twenty-three geostationary satellites per year were scheduled for launch. According to Euroconsult, 89 satellites worth $5.6 billion were launched between 1980 and 1988; for 1989–2000 those numbers would likely rise to 173 satellites worth about $12 billion. The price tag for launching 173 satellites would be about $11 billion.[36] Space may hold even greater economic potential in the future through microgravity manufacturing (of crystals and chemicals, for instance) or even energy production.

European collaboration in space technologies began after purely national space programs proved untenable. The sheer scale of national investment required to join the space race led national policy-makers to turn to cooperation. As Michiel Schwarz asserts, "In the early 1960s, the western European countries were convinced that no individual country was in a position to embark in space programmes on its own at any reasonable cost."[37] In France and Great Britain this realization came only after considerable resources had been devoted to national launcher development. For West Germany, prohibited in the first postwar decade or so from aerospace activities, cooperation offered the fastest way to reenter the field. Perhaps more importantly, collaboration was politically necessary: By tying its space efforts to West European programs, Germany could prove that its ambitions in space were entirely peaceful and cooperative.

36. Patrick Dubarle, "Space: Beginnings of a New Competitive Industry," 12–13; Euroconsult, *World Space Industry Survey: Ten Year Outlook*, 254–59.
37. Michiel Schwarz, "European Policies on Space Science and Technology 1960–1978," 207.

Space collaboration began with the European Launcher Development Organization (ELDO) and the European Space Research Organization (ESRO), both of whose conventions were ratified in 1964.[38] ELDO began with the British offer of the Blue Streak missile (begun as a weapon system and later dropped) to serve as the first stage of a European launcher. The French offered their Coralie rocket as the second stage. Other countries added other ingredients to the stew: The third stage would be German, the test satellite Italian, the telemetry links Dutch, the guidance station Belgian, and the launch site Australian. Passage of the annual ELDO budget required approval by two-thirds of the members, whose contributions had to add up to 85 percent of the total. Oddly, the characteristics of the satellites to be launched on Europa I (the name chosen for the rocket) did not enter into the considerations of ELDO.

For ESRO the initiative came from European scientists (linked to CERN). Its work was limited to space science (that is, research satellites as opposed to applications satellites).[39] The satellites would be launched on National Aeronautics and Space Administration (NASA) or ELDO rockets, if and when ELDO rockets became available. Furthermore, ESRO would have its own facilities. Voting would be by simple majority, each state receiving one vote. Budgets would be approved every three years (as opposed to every year in ELDO) by two-thirds of the members, and the contribution of each state would be proportional to its national income.

By 1968 both organizations were foundering. ELDO was plagued by technical failures: Engineers belatedly discovered that the French and German stages of Europa I would not fit together,[40] and its successor, the Europa II, never achieved a successful launch. Costs were skyrocketing, and some members (first the British and later the Germans) favored buying launch services from NASA. ESRO was riven by *juste retour* disputes and disagreement over whether the organization should enter into the field of applications satellites or remain restricted to scientific satellites.

38. For the history of ELDO and ESRO and of the origins of the ESA, see Schwarz, "European Policies," and Walter A. McDougall, "Space-Age Europe: Gaullism, Euro-Gaullism, and the American Dilemma."

39. CETS was established by the PTTs as a user forum in which the Europeans could band together to deal with Intelsat on communications satellites.

40. Peter Marsh, "How Britain Threw Away Its Lead in Rockets," *Financial Times,* 8 June 1984, p. 22.

ELDO never managed to pull itself together, but ESRO reforms enacted around 1970 were later adopted as the basis for the ESA. The problem of fair return was largely alleviated in ESRO by 1970 because European aerospace companies formed three large, international consortia (Mesh, Star, and Cosmos) to bid for the contracts. In the early 1970s ESRO was awarding contracts such that national shares approximated national contributions to a remarkable degree.[41] The organization was restructured. States would contribute only to those projects they desired at the level they desired (untied to national income).

In order to revive Europe's prospects in space technologies, the European Space Conference was established in 1968, joining ELDO, ESRO, and CETS in a single forum. The aerospace industry argued for increased European cooperation. Indeed, an industry lobby called Eurospace had been formed in 1961 to advance the cause of joint European space efforts. Over five years, through convoluted bargaining, the states participating in the European Space Conference struck deals that led to the ESA. ELDO and ESRO were merged by January 1974. The key bargain was one struck between the French and West German ministers: The French would approve of participation in American post-Apollo activities (Spacelab), and the Germans would approve of the development of a new (French) launcher, the L3S, which later became Ariane. France, the Federal Republic of Germany, and Great Britain each took the lead in a major ESA project, as shown in Table 5.1. All members were required to contribute on a proportional basis to the basic program of scientific research (including running the ESA facilities) but would participate in the optional programs at the level they desired, as in the reformed ESRO. The new ESA took over in their entirety the ESRO programs and facilities.

Since its inception ESA has rung up an impressive number of successes. It has produced successful scientific projects (like the Spacelab and the Giotto craft that passed by Halley's comet in 1986) and applications (in meteorology, earth observation, marine navigation, and telecommunications). Additional satellites were under development in 1986.[42] But the centerpiece of ESA's achievements is the Ariane launcher, which by 1984 had booked a full schedule

41. Schwarz, "European Policies," 212.
42. See ESA, *ESA Annual Report, 1986,* 211.

TABLE 5.1.   CONTRIBUTIONS TO THE EARLY ESA
             PROGRAMS, IN PERCENT

	Ariane Launcher	Spacelab	Marots[a]
France	62.50	10.00	12.50
Germany	20.12	52.55	20.00
United Kingdom	2.47	6.30	58.70
Total[b]	85.09	68.85	91.20

SOURCE: Michiel Schwarz, "European Policies on Space Science and Technology, 1960–1978," *Research Policy* 8 (1979): 226.
[a]Marots was a maritime communications satellite.
[b]Totals sum to less than 100 percent because they represent only the shares of the three countries shown; other countries participated in all three projects.

of satellite launches for three years in advance[43] and by mid-1986 had won half the market for commercial satellite launches, a share worth $1.5 billion in contracts.[44] Furthermore, ESA had begun preparatory programs in several ambitious directions, including a larger model of Ariane (the Ariane 5), a powerful new cryogenic motor for the Ariane 5, a small space shuttle (Hermes), and elements of a manned space station (Columbus).[45]

In at least one important respect ESA differs from Airbus. ESA does not conduct commercial activities. All of its R&D is scientific or, in the applied areas, noncommercial. For instance, the telecommunications satellites (OTS, ECS-1, and ECS-2) have been experimental or demonstrator models, with the leasing of transponders handled through Eutelsat. Commercial satellites are developed outside ESA, as with the collaborative French and German satellites for direct television broadcasting, TDF-1 and TV-SAT. ESA continues to perform the technical development work for the Ariane series, but a separate private venture, Arianespace, handles production of the rockets and marketing for commercial launches. Arianespace is headquartered in France, and its main shareholders

43. See David Marsh, "Ariane Competing with U.S. Technology," *Financial Times*, 23 May 1984, p. 36.
44. David Dickson, "Redesign of Ariane Is Underway," 411.
45. See Pierre Langereux, "L'ESA aura besoin de plus de 200 milliards F d'ici à l'an 2000," 25–28.

are the principal European aerospace and electronics firms, along with some major banks.[46]

What are the organizational features of ESA that might carry lessons for technological collaboration in Europe? First is the variable geometry, or à la carte, nature of participation. ESA's thirteen members[47] are obligated to contribute financially only to the "mandatory" activities, which include the scientific program, the running of ESA's own facilities, and all headquarters functions (documentation, administration of contracts, and the like). Assessments for the mandatory activities are based on national income. For the optional programs, states contribute to and work on only those projects they choose and for the share they desire.[48] The optional programs account for about 80 percent of the total ESA budget, which reached $1.7 billion in 1990. The bulk of the funds go to the major optional programs, including the Ariane rockets, the Hermes space shuttle, and space stations.[49] This arrangement allows states to be selective; so the French shouldered over 60 percent of the costs of developing Ariane,[50] and the Germans paid over half the development costs of Spacelab[51] and committed to 37 percent of the Columbus space-station effort.[52]

A second factor making ESA a success is that it is structured so that countries can obtain a fair share of the contracts for their companies. There are three elements to ESA's solution to the perennial *juste retour* problem. First, the organization is committed to a fair distribution of contracts. The *Convention* declares that the Agency shall "ensure that all Member States participate in an equitable manner, having regard to their financial contribution," and that it need not tender by competitive bidding "where this would be incompatible with other defined objectives of industrial policy."[53] In fact, at a January 1985 meeting ESA ministers passed, at the particular insistence of the small states, a resolution requiring that by

46. Gabriel Lafferranderie, "Les modes de coopération dans le domaine des activités spatiales," 84.

47. Austria, Belgium, Denmark, France, Germany, Ireland, Italy, Netherlands, Norway, Spain, Sweden, Switzerland, and the United Kingdom. Canada and Finland have agreements of association with ESA.

48. *Convention of the European Space Agency,* Articles V.1 and XIII.1–2.

49. ESA, *ESA Annual Report, 1986,* 188; Langereux, "L'ESA aura besoin," 25; Peter Coles, "Is There Profit as Well as Pride?" 140.

50. David Dickson, "Ariane Challenges U.S. on Satellites," 22.

51. Ian Pryke, "Bound for Space," 16.

52. David Dickson, "Europe Plans Its Own Mini Space Station," 816.

53. *Convention of the European Space Agency,* Article VII.1.c–d.

TABLE 5.2.   RATIOS OF CONTRACTS RECEIVED TO
ESA CONTRIBUTIONS

	1 January 1972– 31 December 1986	1 January 1984– 31 December 1986
Austria	0.95	0.84
Belgium	0.91	0.93
Canada	0.89	0.75
Denmark	1.07	1.30
France	1.03	1.02
Germany	1.04	1.00
Ireland	1.22	0.88
Italy	0.95	1.10
Netherlands	0.96	1.16
Norway	0.96	1.03
Spain	0.90	1.06
Sweden	0.94	1.02
Switzerland	0.97	0.80
United Kingdom	1.02	0.90

SOURCE: European Space Agency, Industrial Policy Committee, *Geographical Distribution of Contracts*, ESA/IPC(87)15 (Paris, February 1987), tables 1 and 3.

the end of 1987 all members receive contracts worth at least 95 percent of their contribution to ESA.[54] Table 5.2 shows how closely ESA approximated an ideal return (a ratio of contracts to contributions of 1.0 for each country) over the long run and for 1984–86.

The second part of ESA's response to *juste retour* challenges is that bidding for contracts generally takes place via large consortia involving firms from a number of countries. Two of the old ESRO consortia are still around (Mesh and Cosmos), though most contracting takes place through other groupings, according to one official of the Centre National d'Etudes Spatiales (CNES), France's space agency.[55] Indeed, the old boundary is blurred, MBB (of Cos-

54. Interview 64 (see note about interviews at beginning of Bibliography). ESA, *ESA Annual Report, 1986*, 189.
55. Interview 63.

mos) having merged with ERNO (Mesh). Indeed, each major project is a consortium. For the Columbus project, Dornier of Germany is the lead contractor for the free-flying laboratory module, British Aerospace has the main contract for the polar platform, and Aeritalia heads work on the overall design.[56] Development work for the European Retrievable Carrier unmanned laboratory platform was entrusted to twenty-four European companies headed by MBB/ERNO.[57] Ariane itself is the product of a huge consortium, headed by CNES and involving Aerospatiale, Société Européenne de Propulsion (France), MBB/ERNO, Matra (France), Contraves (Switzerland), and Air Liquide. These principal contractors then subcontract work to companies throughout ESA's membership.[58] Through the consortium and subcontracting system hundreds of European firms participate in every project.

The third element in ESA's success in dealing with the *juste retour* problem is the sheer size of the overall endeavor. ESA has taken on new and ambitious projects, increasing the opportunities for enterprises and hence also the likelihood that each country will be able to land enough interesting pieces. In other words, in a small pie, each slice is a big deal. With a much larger pie, everybody can pick two or three attractive pieces. From 1975 to 1984 ESA spent a total of 870 million UAs (or $710 million). Its total expenditure for 1985–95 was scheduled to rise to 16,684 million UAs but under a new long-term plan proposed by the directorate, that would increase still further to about 20,774 million UAs over 1987–96.[59] It is still not clear whether member governments will approve all the proposed increases. But, as one ESA official told me, there is enough work to go around.

CONCLUSIONS

At the end of this diversion into aerospace collaboration, what analytical points will be useful in thinking about telematics collaboration? Table 5.3 summarizes the main variables proposed in the analytical framework.

56. Dickson, "Europe Plans," 16.
57. "EURECA Industrial Contract Signed," 429.
58. Lafferranderie, "Les modes de coopération," 84.
59. Lafferranderie, "Les modes de coopération," 80; Langereux, "L'ESA aura besoin," 25.

TABLE 5.3.   NECESSARY CONDITIONS FOR
INTERNATIONAL COLLABORATION

	EC	Unidata	Airbus	ESA
Cognitive change				
Unilateral strategies failed	N	N	Y	Y
States in policy adaptation	N	N	Y	Y
Political leadership				
From major states	N	N	Y	Y
From an IO	Y	N	N	Y
Arrangements for *juste retour*				
A la carte	P	Y	N	Y
Authoritative allocation	N	N	Y	Y

Notes: The ambitious science and technology schemes of Spinelli were never adopted. The EC programs that were approved (COST, CREST) were à la carte.
Y = condition present
N = condition not present
P = condition partially present

In striking contrast to the telematics sectors during the 1970s, two crucial conditions were present for aircraft and space. First, purely national policies had failed to create viable capabilities in passenger aircraft and space. This failure led to policy crisis; there was a perception in each of the major states that they could not afford to be excluded from the civilian aircraft industry or from space systems. Policy-makers were therefore prepared to adapt their approaches, as evidenced by the international discussions initiated in the early 1960s for the purpose of combining launcher efforts in ELDO, and in 1966 for the creation of a European aircraft. Governments were in an adaptive mode.

Second, in each case, a political leader stepped forward to pay the costs of organizing collaboration. In the case of space IOs provided a forum for reaching the bargains that created the ESA, and the drive to reach agreement came from among the members of those organizations. A particular enthusiasm for European space science and technology motivated France to play a leading role in mobilizing support for a new space organization. The Airbus bargain, in contrast, was intergovernmental, with no participation by an IO (despite attempts by the Commission to catalyze European cooperation). The leading states were France and Germany, with French aerospace advocates pulling the effort through the difficult initial stage. The point is that leadership is necessary. Major states will frequently assume the entrepreneurial role, but IOs can as well, as later cases will show. By the same token, IOs cannot mobilize cooperation when states do not perceive unilateral strategies to have failed, as the telematics cases in this chapter demonstrate.

In their organizational arrangements for allocating benefits, Airbus and ESA illustrate different approaches. In Airbus, shares of each model are agreed at the outset and fixed by formal agreement. ESA, in the optional programs that account for the lion's share of the budget, utilizes a pure à la carte system—states pick and choose from a large menu. Of course, the two types of arrangement are not mutually exclusive; in ESA, participation in the relatively small scientific programs is mandatory, with each country receiving at least 95 percent of its contribution in work. And, in Airbus, shares can shift in the long run as partners seek to change their level of participation in new models.

Industry plays a crucial role in both the aerospace programs. Indeed, companies and laboratories do the actual bidding for shares of projects. The technical goals and tasks, in fact, are determined largely by industrial participants. Transnational consortia of major companies contribute to the success of aerospace collaboration— as they do also to ESPRIT, RACE, and EUREKA. In addition, both programs are large enough that contracts and subcontracts can be spread across Europe. Consortia involving dozens of participants from many countries bid for and receive the ESA contracts. For Airbus, components and subassemblies flow from throughout the region to the main contractors. In short, there is enough work to go around, and it can be divided and subdivided so as to provide each state an acceptable level of participation.

States turn to collaboration only after national strategies have fallen short. Failure triggers adaptation. The period of national-champion strategies (the 1970s) ended in a crisis for European telematics industries and policies; collaboration emerged in the 1980s. The next chapter uncovers the roots of that crisis and the technological factors that began to channel the adaptive responses toward collaboration.

SIX

When Champions
Aren't Enough

Europe by 1980 was a land of giants as far as telematics is concerned. The hopes of governments for domestic telematics industries rested on champions whose ordination I described in Chapter 4. But the local favorites were not the only titans doing battle for European markets. American and Japanese multinationals exported to the Old Continent and also established their own factories or joint ventures on European soil. By 1980 the Japanese and Americans dominated both the European and world markets for semiconductors and computers, and were cutting deeply into Europe's traditional trade surplus in telecommunications. Furthermore, the scale and pace of U.S. and Japanese R&D threatened to increase the technical advantages of Europe's rivals. These factors amounted to an equation for crisis in Europe.

The continued (and increasing) dominance of American and Japanese telematics firms signaled to many Europeans the failure of traditional policies. A generalized sense of crisis arose, with old approaches discredited and no fresh strategy to take their place. The early 1980s thus saw a technology-gap scare similar to the crisis of the mid-1960s. The feeling of falling steadily and perhaps irretrievably behind the United States and Japan recalled the panic and even much of the rhetoric of the earlier scare.[1] The difference the second

1. See, for example, Margaret Sharp and Claire Shearman, *European Technological Collaboration*, chap. 1; and Andrew J. Pierre, ed., *A High Technology Gap?*

time around was that European leaders added a collaborative element to their policy response.

Japan's emergence constituted the major structural change at the world level in telematics. Europeans had been accustomed to American dominance of electronics markets. Japan's successful challenge to U.S. preeminence provided an additional catalyst to the emerging telematics crisis in Europe. The crisis was aggravated by American and Japanese programs that threatened to shove Europe even further to the rear.

But relative international weakness cannot explain why the European countries collaborated. The Europeans had always been weak in telematics relative to their international competitors yet had never collaborated before. The international setting defined the problem facing European governments but did not (and could not) determine the nature of the European response. That collaboration emerged was due to entirely different factors.

This chapter has two major parts. In the first I briefly describe Europe's international position in the telematics sectors circa 1980, as well as American and Japanese telematics initiatives. I then analyze the technological changes that were responsible for the emergence of a powerful consensus in favor of collaboration among Europe's largest telematics firms. The telematics industry played a crucial role in convincing governments of the importance of ESPRIT and RACE.

THE INTERNATIONAL SETTING

Painting a picture of Europe's troubles and Japan's good fortune will be a matter of numbers and anecdotes. The data show how market shares shifted; the anecdotes illustrate the fate of specific firms.

L'Europe Couchante

As pointed out in Chapter 4, European components makers (and governments) failed to appreciate the importance of digital, large-scale ICs. During the 1970s, digital ICs, both memory chips and microprocessors, became the essential raw material for data-processing, telecommunications, industrial automation, and military and

TABLE 6.1.   SHARES OF THE EUROPEAN MARKET
FOR ICs, 1977–81, IN PERCENT

Firms Based in	1977	1978	1979	1980	1981
Europe	33	33	32	30	30
United States	64	63	63	64	62
Japan	3	4	5	6	8

SOURCE: Organization for Economic Cooperation and Development, *The Semiconductor Industry: Trade Related Issues* (Paris, 1985), 118.

consumer electronics. European producers, formerly strong in discrete and analog IC devices, steadily lost market shares. The share of West European firms in world semiconductor markets fell from about 16 percent in 1978 to about 12 percent in 1983; by 1988 it was down to 10 percent, though it rose slightly in 1990 to almost 11 percent.[2] Europe's performance in ICs, the crucial category of semiconductors, was even more dismal. By 1978 the share of Western Europe in world production of ICs was only 6.7 percent; it declined to 5.8 percent in 1980 and rose slightly to 5.9 percent in 1982.[3] Even in the European market American companies dominated, as shown in Table 6.1.

In addition growth rates for demand and production of ICs in Europe were well below the rates for the United States and Japan. While demand expanded at 20 percent per year in the United States and 19 percent per year in Japan over the period 1978–82, it grew by only 13 percent per annum in Europe. Production also increased more rapidly in the United States (17 percent annually) and Japan (25 percent annually) than in Europe (12 percent per year) during the same stretch.[4] Thus, Europe was not benefiting from the microelectronics revolution in the same way that its trade rivals were. One revealing indicator of this discrepancy is the per capita consumption of semiconductors. In 1984 the consumption of semiconductors in the United States had a value of $52 per capita, and in

2. Jonathan Weber, "U.S. Gains Ground in World Chip Market," *Los Angeles Times*, 3 January 1991, p. D1.

3. OECD, *Semiconductor Industry*, 102.

4. Ibid., 103.

Japan $61 per capita. The average for all Europe was $14 per capita, with Germany marking the high end at $22 per capita.[5] Not only was Europe behind, but its competitors were accelerating faster.

The situation was so bad that even by the mid-1970s there was not a single European producer of standard ICs.[6] Some firms were successful in niche markets, like Ferranti in uncommitted gate arrays. Most European producers built custom or semicustom chips largely for use in their own final products (computers, telecommunications systems, consumer electronics). The IC divisions of the electronics champions were all losing money: In 1980 Siemens had not shown a profit in ICs since 1965 and SGS-Ates never had.[7] Thomson was a consistent loser through 1984.[8] Philips showed a profit on ICs only in 1979, with help from the American firm Signetics, which it purchased in 1975. About half of Philips's IC production went to in-house uses. Philips has been the only European IC maker big enough to make the world top ten, and it slipped from fourth place in 1979 to sixth place in 1983.[9]

The situation in computers was no better. Table 6.2 shows the world's top twenty-five firms in the data-processing industry for 1978, 1983, and 1986. The European computer makers hover near the middle of the top twenty-five, with Siemens and Olivetti moving up and finally cracking the top ten. Broken down by category, the picture is no better. Only one European maker cracked the 1983 top ten in mainframes (Siemens in eighth place), and its two top-of-the-line models had been manufactured under license from Fujitsu since 1978.[10] ICL sold Fujitsu's Atlas 10 (IBM-compatible) mainframe until 1984, and after that its advanced Series 39 contained forty-three chips developed by Fujitsu under a technology agreement. ICL's microcomputers have been Sun workstations.[11] In minicomputers only Olivetti cracked the top ten from Europe (in

5. Michael G. Borrus, *Competing for Control,* 199.
6. See Malerba, *Semiconductor Business,* 119. Standard ICs are commodity ICs sold on the world market as opposed to custom or semicustom chips (sometimes called application-specific integrated circuits, or ASICs).
7. Ibid., 166, 171.
8. Guy de Jonquieres and Paul Betts, "The Euphoria Is Over," *Financial Times,* 7 February 1985, p. 14; Paul Betts, "Thomson Has Another Try," *Financial Times,* 25 October 1985, p. 18.
9. Malerba, *Semiconductor Business,* 164; OECD, *Semiconductor Industry,* 116.
10. Pamela Archbold and John Verity, "A Global Industry: The Datamation 100," 38; Laurence P. Solomon, "The Top Foreign Contenders," 81.
11. Kelly, *British Computer Industry,* 47; Guy de Jonquieres, "Electronics in Europe," *Financial Times,* 28 March 1984, Survey, p. 1.

seventh place), and in microcomputers Olivetti was again the only European member of the top ten, in ninth place.[12]

European firms had not been able to challenge IBM's dominance. IBM as of 1975 held over half the computer market in France, Germany, and Italy, and 40 percent of the market in the United Kingdom.[13] Even in Britain by 1985 IBM mainframes installed outnumbered those of the national champion, ICL, and were selling faster.[14] Indeed, in 1983 IBM Europe had data-processing revenues more than seven times greater than those of its nearest rival, Bull. And Bull was losing money: It showed losses of FFr 1.35 billion in 1982, FFr 625 million in 1983, and FFr 489 million in 1984.[15] In 1983, nine of the top fifteen computer makers in Europe were American firms. IBM alone had data-processing revenues greater than those of the next nine largest manufacturers combined. IBM's share was 42 percent, up from 38 percent in 1981. Of the top twenty-five firms operating in Europe in 1983, thirteen were American.[16] Their combined share of the European market was 81 percent.[17] All this American dominance had happened despite government subsidies and protected markets for the national champions.

Contrary to the situation in semiconductors and computers, in telecommunications in the early 1980s Europe was not suffering from obvious and longstanding failings. In fact, European telecommunications technology was among the most advanced; the French developed the first fully digital switching system in the mid-1970s. Furthermore, none of the European countries with indigenous telecoms-equipment production showed a trade deficit in the sector, as shown in Table 6.3. The EC countries as a bloc managed a telecommunications-equipment trade surplus with the rest of the world of about $1.7 billion in 1982. In other words, the telecoms sector in the early 1980s did not appear to be crying out for help.

But prying into the statistics shows that the rosy overall picture was deceptive. The strong EC trade surplus was built on exports to Third World countries, especially to members of the Organization

12. Archbold and Verity, "A Global Industry," 38.
13. Malerba, *Semiconductor Business,* 181.
14. Kelly, *British Computer Industry,* 15.
15. Guy de Jonquieres and Paul Betts, "The Euphoria Is Over," *Financial Times,* 7 February 1985, p. 14; Guy de Jonquieres, "The Harsh Imperatives of Survival," *Financial Times,* 24 June 1985, Survey, p. 16.
16. Guy de Jonquieres, "Bull Emerges as Biggest European Computer Maker," *Financial Times,* 17 August 1984, p. 6.
17. Malerba, *Semiconductor Business,* 181.

TABLE 6.2. TOP TWENTY-FIVE COMPANIES IN THE
WORLD COMPUTER INDUSTRY, RANKED BY DATA-
PROCESSING REVENUES, 1978–86
(in million current U.S. dollars)

Rank	1978		1983		1988	
	Company	Revenues	Company	Revenues	Company	Revenues
1	IBM	17,072	IBM	36,503	IBM	55,003
2	Burroughs	2,107	DEC	4,827	DEC	12,284
3	NCR	1,932	Burroughs	4,000	*Fujitsu*	10,999
4	Control Data	1,867	Control Data	3,508	NEC	10,475
5	*Hitachi*	1,830	NCR	3,333	Unisys[a]	9,100
6	Sperry Rand	1,807	Sperry[b]	3,072	*Hitachi*	8,248
7	*Toshiba*	1,633	*Fujitsu*	2,800	Hewlett-Packard	6,300
8	DEC	1,437	Hewlett-Packard	2,496	**Siemens**	5,951
9	Honeywell[c]	1,294	NEC	2,299	**Olivetti**	5,428
10	*Fujitsu*	1,248	**Siemens**	2,189	NCR	5,324
11	**CII-HB**	1,061	**Olivetti**	1,816	**Bull**	5,296
12	ICL	1,019	**Wang**	1,793	Apple	4,434

13	**Olivetti**	789	*Hitachi*	1,700	*Toshiba*	4,226
14	**Siemens**	703	Honeywell	1,666	*Matsushita*	3,441
15	*NEC*	672	**Bull**	1,527	*Canon*	3,391
16	Hewlett-Packard	657	**ICL**	1,283	Control Data	3,254
17	**Philips**	602	Xerox	1,156	Wang	3,074
18	Memorex	570	**Philips**	1,095	**Nixdorf**	3,044
19	**Nixdorf**	554	Apple	1,085	**Philips**	2,794
20	Itel	487	AT&T	1,080	Xerox	2,650
21	TRW	466	**Nixdorf**	1,063	AT&T	2,445
22	Data General	380	TRW	1,015	**STC**[d]	2,425
23	Amdahl	321	**Ericsson**	971	Memorex Telex	2,078
24	Storage Technology	300	Tandy	945	Compaq	2,065
25	Automatic Data Processing	290	Commodore	927	*Nihon Unisys*	2,057

SOURCES: 1978: Becky Barna, "The Datamation 50," *Datamation* 25 (25 May 1979):18; Laurence P. Solomon, "The Top Foreign Contenders," *Datamation* 25 (25 May 1979):79. 1983: Pamela Archbold and John Verity, "A Global Industry: The Datamation 100," *Datamation* 31 (1 June 1985):50. 1988: Joseph Kelly, "The Datamation 100: Three Markets Shape One Industry," *Datamation* 35 (15 June 1989):11.

Notes: The names of Japanese companies are in italics; those of European companies are in boldface.

[a] Unisys was created by the merger of Sperry and Burroughs in 1986.
[b] "Rand" was dropped from the company name.
[c] Does not include CII-HB
[d] STC bought ICL in 1984.

TABLE 6.3.   TRADE BALANCES IN
TELECOMMUNICATIONS EQUIPMENT, 1980
(in million U.S. dollars)

	Exports	Imports	Balance
Germany	2,222	1,204	1,018
Sweden	993	233	760
Netherlands	1,269	742	527
United Kingdom	1,240	778	462
France	1,154	781	373
Belgium-Luxembourg	725	515	210
Switzerland	285	250	35
Italy	596	593	3
Japan	3,772	209	3,563
United States	2,655	3,212	−557

SOURCE: Organization for Economic Cooperation and Development, *Telecommunications: Pressures and Policies for Change* (Paris, 1983), 133.

of Petroleum Exporting Countries (OPEC). In fact, about 80 percent of equipment exports in 1980 (not counting trade within the EC) went to the Third World, as seen in Table 6.4. The EEC trade surplus in telecoms equipment began declining in the early 1980s; 1985 was the third straight year of contraction, with the surplus dropping to 1,247 MECU from 1,533 MECU the year before. In addition the EC registered a growing trade deficit in the sector vis-à-vis the United States and Japan. Its deficit with the United States grew 25 percent in 1985 to 657 MECU, and that with Japan rose by 61 percent to 582 MECU.[18] These trade deficits with the United States and Japan suggested that Europe was weak in the most advanced sectors of the telecommunications market.[19]

Certainly Europe has been slower than the United States and Japan in the diffusion of new products and services. This lag is due in part to the traditional role of the PTTs, which have monopolized the provision of networks (except in the United Kingdom), limited the provision of new services (like VANs), and until recently con-

18. CEC, *Towards a Dynamic European Economy*, 158.
19. Borrus et al., *Telecommunications Development*, 38.

TABLE 6.4.   TELECOMMUNICATIONS-EQUIPMENT
EXPORTS BY REGION AND DESTINATION, 1980
(in million U.S. dollars)

			Destination			
From	*EC*	*Japan*	*United States*	*EFTA*	*OPEC*	*Other*
EC	2,485	40	199	733	1,348	2,723
Japan	785		1,096	113	253	1,434
United States	714	123		115	257	1,209

SOURCE: Organization for Economic Cooperation and Development, *Telecommu-nications: Pressures and Policies for Change* (Paris, 1983), 135.

trolled the kinds of terminal equipment that could be attached to the network. Thus, for example, on-line data services in Europe were worth about $200 million in 1982 as compared with $800 million in the United States. In videotex-type services (interactive on-line data banks), there were more subscribers in the United States than in all the EC. Europe lagged behind the United States in commercial satellites by about ten years; the marketing and promotional budget of just one American business satellite system, Satellite Business Systems, at $200 million, surpassed the entire EC investment in such services.[20] Facsimile, an advanced document-transmission technology, spread in Europe more slowly than in the United States: While the United States had over 225,000 terminals installed in 1980, Europe could count only 47,400.[21] All these signs led European telecommunications administrators, suppliers, and their governments to sense a dangerous weakening in Europe's traditional electronics stronghold.

Le Japon Levant

Some of the tables in the preceding section depict the broad lines of Japan's accelerated rise in advanced electronics. But additional details will show more clearly why Japan's surge was so frightening

20. As estimated by A. D. Little, *European Telecommunications: Strategic Issues and Opportunities for the Decade Ahead, Executive Report,* 25, 29.
21. van Tulder and Junne, *European Multinationals,* 33.

to the Europeans. Remember that even in the 1960s Europe may have been slightly ahead of the Japanese in telematics because European scientists and enterprises pioneered many of the early advances in solid-state physics and semiconductors. Japan, following the guidance of MITI, first built its way to world leadership in steel and shipbuilding. Its entrée into electronics was consumer products—transistor radios, calculators, and televisions. By the 1980s Japanese producers dominated world markets for videotape recorders and compact-disc players (the high end of the consumer-electronics spectrum). Consumer products provided the initial demand for Japanese semiconductor companies, but computers and industrial applications rapidly became the chief sources of demand pull for advanced ICs in Japan.[22]

Japan overtook the United States first in standard memory chips. Although a technology follower in the first generations of RAM circuits, Japan developed and produced 64K DRAMs (dynamic RAMs) ahead of American companies.[23] Whereas Japanese firms captured only 12 percent of the world market for 4K DRAMs in the mid-1970s, they took 70 percent of the world market for 64K chips in 1981 and 90 percent for 256K DRAMs in 1984.[24] Japan's trade balance in ICs converted from a $142 million deficit in 1976 to a $239 million surplus in 1981.[25] Although the Japanese share of the European IC market remained below 10 percent, it was growing. Japanese firms exported ICs worth only $12 million in 1976; that value rose to $165 million in 1980.[26]

Japan eventually surpassed the United States in the RAM market. In fact, by 1985 all but two American merchant producers of DRAMS had been driven from the market (Texas Instruments and tiny Micron Technology remained). Japanese firms were the first to produce one-megabit memory chips (with four times as much memory capacity as a 256K chip) and four-megabit circuits. In 1985, for the first time, a Japanese company, Nippon Electric Company (NEC), was the world's largest producer of semiconductors.[27] A year later

22. Malerba, *Semiconductor Business,* 176.
23. Borrus, *Competing for Control,* 143.
24. Malerba, *Semiconductor Business,* 155.
25. Ibid., 156.
26. Ibid., 139.
27. Michael Feibus, "Chip Companies Look for a Second Good Year," *San Jose Mercury News,* 4 January 1988, p. C2.

Japan's total world market share for ICs surpassed that of the United States for the first time, at just over 45 percent.[28] Furthermore, by 1986 Japanese companies were beginning to challenge U.S. dominance of the microprocessor realm.[29]

The story is not quite as dramatic in computers and telecommunications equipment, but Japanese advantages in IC production were increasingly conferring advantages on Japanese systems producers. In fact, the major Japanese semiconductor makers (NEC, Toshiba, Hitachi, Fujitsu, Mitsubishi, and Matsushita) all sold computers and telecommunications systems as well. As seen in Table 6.2, the number of Japanese computer firms in the world's top twenty-five rose from four in 1978 to six in 1986. Furthermore, Japanese companies like Fujitsu and Hitachi were providing computer technology (and sometimes the machines themselves) to European producers like Siemens, Badische Anilin-und Sodafabrik (BASF) and ICL. In telecommunications equipment the share of European manufacturers in total world exports in 1983 had been contracting at a rate of 1 percent per year for ten years, while the Japanese share had been increasing at a similar rate.[30] EEC imports of Japanese telecommunications equipment did not constitute a large share of the market, but they had been growing steadily. The Commission lamented in its Green Paper on telecoms that imports from Japan in 1985 totaled 616 MECU, while exports to Japan totaled only 34 MECU.[31] Of course, this discrepancy is as much a trade problem as a technology problem, but the Europeans had much reason to be nervous about the technological side also, as we shall see later in this chapter.

Striking as they are, the figures on Japan's telematics surge tell only half the story. Europeans were concerned not solely with the data but with how Japan achieved its breakthroughs. A set of government institutions and policies were behind the Japanese miracle. Europeans feared that the Japanese state would continue to force the technology pace, propelling the country ever further ahead. Although scholars have written shelves of books on Japanese indus-

28. John Burgess, "U.S. Seeks Silicon Island Beachhead," *International Herald Tribune*, 1 April 1987.
29. Borrus, *Competing for Control*, 176–77.
30. A. D. Little, *European Telecommunications*, 39.
31. CEC, *Towards a Dynamic European Economy*, 167.

trial policies, it is possible to summarize succinctly the principal features of the Japanese system that have encouraged accelerated growth in the telematics sectors.

Japan has confounded conventional economic wisdom by channeling economic and technological resources into sectors chosen by government agencies as being conducive to long-term growth.[32] There have been three principal elements of the Japanese approach:

First, *targeting technologies* is the province of MITI, which selects those industries that are likely to experience rapid technological change and world market growth over the long run. MITI accomplishes this task through constant consultations with industry and university scientists. MITI had already in the late 1960s selected semiconductors and computers as the sectors with which to improve Japan's competitiveness.[33]

Second, MITI and other government agencies became gatekeepers, *protecting the home market* for key sectors. Tariffs and quotas limited foreign access to markets where Japanese firms were trying to establish themselves, like semiconductors and computers. The government also reviewed all applications for foreign investment in Japan; foreign companies could establish themselves within the Japanese market only as junior partners in joint ventures with domestic firms and only by granting Japanese firms access to advanced technologies. Thus, in the period when American firms dominated world semiconductor and computer markets, only Texas Instruments and IBM could establish subsidiaries in Japan.[34]

A third and crucial part of the Japanese story has been *government-sponsored, state-of-the-art R&D* programs, which diffuse results among the principal firms. Two essential features of Japanese technology-development programs have been their cooperative character and government funding. For instance, the first major telematics R&D program aimed at producing the technologies to compete with IBM's 360 series. All the major electronics firms participated in the research and all had equal access to resulting patents

32. In the account that follows, for general Japanese practices I rely on Freeman, *Technology Policy,* especially 33–49; and Dosi, Tyson, and Zysman, "Trade, Technologies, and Development." For specifics about Japanese promotion of the semiconductor and computer industries I rely on Borrus, *Competing for Control;* and Marie Anchordoguy, "Mastering the Market: Japanese Government Targeting of the Computer Industry," 509–43.

33. Borrus, *Competing for Control,* 119–27.

34. Ibid., 119–21; Anchordoguy, "Mastering the Market," 513–17.

(which the government owned). The New Series program launched in 1971 aimed at developing large-scale integration components and the production technologies needed to compete with IBM's 370 series. MITI financed joint R&D (to the tune of hundreds of millions of dollars) and oversaw the formation of industry groupings. NTT (the government telecommunications authority) supervised much of the research.

In 1976 MITI, NTT, and the five largest semiconductor/computer companies joined in setting up the VLSI program. Again, the companies collaborated on R&D, with government funding of about $121 million. The VLSI program produced the breakthroughs that sent Japanese firms into the lead in ICs and allowed them to produce IBM-compatible computers that outperformed IBM. Japanese firms derived advantages from economizing on R&D by cooperating on generic technologies, then competing among themselves in the protected Japanese market.[35]

In short, the problem for Europe was not just that Japan had grown but how it had grown. The combination of government policies and practices encouraged a formidable system of innovation that promised to propel Japan a generation ahead of the competition, leaving Europe to glean the kernels left behind by Japan's fast-moving reaper.

Un Monde Menaçant

A final element in the telematics world that threatened further erosion of Europe's position was that both the United States and Japan had in place fresh programs designed to benefit their enterprises. I will describe first U.S. and Japanese initiatives in the semiconductor and computer fields, then those in telecommunications.

In the United States the assault on advanced microelectronics and computing came from the Department of Defense. Given the U.S. military's increasing reliance on high-tech weaponry, possessing leading-edge technologies had become a matter of national security. This was the logic behind the very-high-speed integrated-circuit (VHSIC) program and the Strategic Computing Initiative (SCI). The VHSIC program, launched in 1980, aimed at developing components comparable in level of integration (the number of electronic

35. Borrus, *Competing for Control,* 126–27; Anchordoguy, "Mastering the Market," 526–30.

elements packed onto the chip) to next-generation commercial ICs. But VHSIC chips also had to meet military requirements for low power consumption, low maintenance demands, built-in testing, durability, and radiation hardening. The R&D would be carried out by American companies selected by competitive bidding. For the first two phases firms on the VHSIC roster included such American telematics stars as IBM, TRW, Motorola, Sperry, Signetics, Burroughs, Texas Instruments, National Semiconductor, CDC, Honeywell, Fairchild, Westinghouse, and Hewlett-Packard. The total program received a budget of $680 million over eight years, on top of regular Department of Defense spending on electronics R&D.[36]

In October 1983 the Pentagon's Defense Advanced Research Projects Agency announced a new program designed to make breakthroughs in advanced computing and artificial intelligence (AI). SCI R&D covers microelectronics, hardware and software architectures, intelligent functions (reasoning, vision, speech recognition), and military applications (like an autonomous land vehicle for the Army and a "pilot's associate" for the Air Force). Initial funding for SCI was $600 million for the first five years.[37]

Of course, the granddaddy of all American high-tech defense projects is SDI, or Star Wars. With an initial projected budget of $26 billion, SDI promised to advance the state of the art in numerous fields: lasers, new materials, communications, AI, particle beams, and others. As with the other major U.S. technology programs, SDI's impacts on civilian industry were ambiguous. Within the United States intense debate arose over whether SDI (and the VHSIC program and SCI) would help or handicap American industry in commercial competition.[38] In Europe the virtually universal perception was that the vast sums of money involved and the ambitious research agenda would push U.S. contractors to breakthroughs that would entail commercial advantages. American defense R&D programs in the 1980s reminded Europeans of the earlier NASA and Air Force programs that made possible the early American blastoff in ICs and computers. Indeed, as Chapter 9 will

36. This description of the VHSIC program comes from Leslie Brueckner, with Michael G. Borrus, *Assessing the Commercial Impact of the VHSIC Program,* 8–14.
    37. Dwight B. Davis, "Assessing the Strategic Computing Initiative," 41–49.
    38. See Brueckner, *Assessing the Commercial Impact;* Jay Stowsky, *Beating Our Plowshares into Double-Edged Swords;* and Davis, "Assessing the Strategic Computing Initiative."

show, EUREKA was triggered by SDI and the fears it inspired among European decision-makers.

The Japanese deployed an impressive array of programs designed to push the technology frontier. Among these were the programs in optoelectronic components, budgeted at $77.5 million for the 1980s, and in New Function Elements, with $100 million for the period 1982–89.[39] The New Function Elements program was aimed at next-generation components, including super-lattice and three-dimensional elements and hardened (radiation-proof) circuits. A further project targeted supercomputers, with $92.3 million over 1982–90. But the most impressive, and to the Europeans the most frightening, program was the so-called Fifth Generation Computer (5G) project.

The 5G project, begun in 1982, aims at developing the software and hardware technologies needed for the next generation of computers. With 5G Japan hopes to leapfrog into the lead in AI. This goal will require technologies enabling computers to "think"—that is, to manipulate information via rules of logic (as opposed to simply carrying out instructions to perform mathematical operations). Like the VLSI program, 5G is run by MITI at a laboratory dedicated to that project with participation by NTT and the six principal Japanese computer firms (non-Japanese researchers were invited to participate but in practice 5G is virtually all Japanese). The original budget for 5G was about $400 million over ten years, though actual spending will fall short of that level. Spending for the first three-year phase totaled $33.7 million, somewhat below the $40 million originally foreseen.[40] Still, the 5G program alone provoked a series of imitator programs in Europe, as we shall see in the next chapter.

In the telecommunications realm developments in the United States and Japan caused serious consternation in Europe. The American telecommunications system coalesced in 1934 with the creation of the Federal Communications Commission (FCC) and the recognition of a natural monopoly in the provision of telecommunications, namely, AT&T's. Having bought up most of the small private telephone companies across the nation, AT&T would thereafter offer universal service, under the regulatory eye of the FCC. The mo-

39. OECD, *Semiconductor Industry,* 79.
40. Tom Manuel, "Cautiously Optimistic Tone Set for 5th Generation," 57–58.

nopoly extended beyond networks, as AT&T owned its own R&D facilities in Bell Labs, as well as a manufacturing arm, Western Electric, to supply the system. However, AT&T was prohibited from competing overseas; its foreign subsidiaries had mostly been bought by a company that called itself ITT. This was the arrangement that would, over several decades, be deregulated.

The government brought an antitrust suit against AT&T in 1954. The consent decree agreed to by the parties meant that AT&T kept its U.S. monopoly, but Bell Labs was required to make available its patents, and Western Electric could not sell equipment abroad. AT&T could not enter computer markets, though it possessed the technologies and the resources to do so. In 1959 the FCC allowed the creation of microwave links for data communications within companies. The FCC began in 1966 a far-ranging inquiry (now called Computer I) into data communications and the network and service monopolies of AT&T generally. The Carterphone decision in 1968 permitted the attachment of non-AT&T terminal equipment to the network. An FCC decision in 1971 allowed other companies to offer data-transmission services. In 1976 the FCC made it possible for specialized carriers to resell network capacity leased from AT&T; this decision in effect meant that competitors could offer both basic telephone and data services fully connected to the AT&T system. Companies like MCI and Sprint began to compete for the provision of basic long-distance services. Yet AT&T could offer only basic telephony, even though its competitors could provide all services.

Finally, in 1982 a second antitrust case and the Computer II inquiry concluded with AT&T and the Justice Department agreeing to a modified final judgment (it modified the 1954 consent decree). Under its terms AT&T could compete in enhanced services after divesting itself of the local Bell operating companies (BOCs). The twenty-two local BOCs were hived off to seven regional holding companies, which could offer both basic and enhanced services under FCC regulation. AT&T retained only its interexchange (long-distance) lines. More important for world competition, AT&T could enter the computer market and sell equipment overseas. The breakup of AT&T constituted the deregulatory volley that was heard around the world.

Simultaneously, the Department of Justice settled its longstanding antitrust suit against IBM, permitting the computer colossus to enter telecommunications markets, from which it had been blocked.

Thus, as of 1984 the two giants AT&T and IBM would each attack the other's base markets—AT&T entering computers and IBM, telecoms.[41] Furthermore, the world would be the battleground, and because Japan constituted a fairly closed market, that meant Europe. The fears aroused in Europe by that prospect were reaching their peak precisely during the years RACE was getting underway, namely, 1984–85. The American threat therefore figured prominently in the selling of RACE, as I will show in Chapter 8.

Events seemed to confirm many European apprehensions. AT&T moved quickly into Europe by forging alliances with European companies. First came a joint venture with Philips, AT&T-Philips Telecommunications, to develop and market digital exchanges. Then AT&T bought a 25 percent stake in Olivetti. Since 1984 AT&T has not succeeded in selling its ESS-5 digital switch in Europe (winning orders only in the Netherlands). But in the initial postdivestiture period, the threat was real. No one could have known then that AT&T would fail to execute a coherent strategy.

IBM also opened its offensive in Europe; IBM was not interested in public switches but in other network equipment (like PABXs) and especially in hardware and software for VANs.[42] IBM vigorously sought a contract with the Bundespost to develop and supply its videotex system, Bildschirmtext. IBM won the contract, though

41. The IBM ventures into telecommunications have not panned out. In fact, by 1989 IBM had washed its hands of both Satellite Business Systems (satellite communications networks) and Rolm (PABXs). Nevertheless, at the time, the IBM moves appeared as genuine threats in Europe and were constantly referred to as such.

42. A note on terminology is appropriate here. VANs stands for value-added networks, also called value-added services. The term implies a difference from basic services. Basic services are those involving the transmission of voice or nonvoice information, without changing or manipulating the information. Voice telephone, telex, and facsimile services fall in the basic-services category. Telex and teletext services constitute a gray area; if no additional services besides transmission (such as storing or forwarding) are performed, they are basic services. With storing or forwarding, telex and teletext could be considered value-added services. PTTs wishing to protect their monopolies prefer to call them basic services, even with the additional operations. Enhanced services are those in which something in addition to mere transmission is provided, "when the information provided by the sender is changed, stored, manipulated or otherwise acted upon in the network." Within enhanced services Aronson and Cowhey distinguish information services from VANs. Information services include databases, data-processing services, and on-line services. VANs include protocol conversion for linking terminals, packet assembly and switching, storage and forwarding of messages. For simplicity, I will lump information services and value-added services together under the label *VANs*. These definitions and quotes come from Jonathan David Aronson and Peter F. Cowhey, *When Countries Talk: International Trade in Telecommunications Services*, 85–99.

reportedly by bidding so low that it has not been profitable.[43] A similar arrangement with British Telecom was nixed at the last minute by the British agency regulating mergers. IBM created a joint venture with Fiat-Telettra (1986) to develop data networks but only after IBM had failed to land an agreement for establishing a data network with Societa Italiana per l'Esercizio delle Telecomunicazioni (SIP), the concession operating 80 percent of Italy's networks.

The American approach to deregulation has exerted pressure on traditional telecoms institutions abroad in other important ways. One result of deregulation has been that American users, especially the large corporations, have had access to new equipment, advanced services, and competing networks. In fact, the largest corporate users today, like Hewlett-Packard and McKesson, build and run private telecoms networks, complete with their own transmission facilities and switches. Most businesses employ a combination of private and public telecommunications networks. Thus, U.S. businesses have benefited from better telecoms facilities at lower prices than their European counterparts. Lack of access to advanced services and higher telecommunications costs overall led European businesses to favor Commission plans for coordinated, Europe-wide modernization and liberalization.

Japan also posed challenges to Europe's fragmented telecommunications systems. The Japanese had ambitious plans for the transition to broadband networks and had also altered the traditional PTT arrangement. A new telecommunications law in April 1985 provided for the gradual privatization of up to 49 percent of the shares in NTT, the government remaining the chief shareholder. Under the new regulation NTT had to compete with other companies for the provision of both basic and enhanced services. The former monopoly provider of international telecommunications services, Kokusai Denshin Denwa (KDD), also had to face competition from a rival company authorized by the ministry of posts and telecommunications.

Beyond these liberalization measures Japan planned an aggressive drive into future broadband communications. NTT was to be the engine for the development of the Information Network System, aimed at providing the hardware, software, and services to carry simultaneously voice, data, audio, and visual signals to all subscrib-

43. Hart, "The Politics of Global Competition," 186.

ers, not just the major companies in dense urban areas. The technologies needed were to be developed by NTT with its family of
traditional suppliers (NEC, Fujitsu, Hitachi, and Oki), following
the pattern of successful Japanese technology programs of the past,
like VLSI. In addition NTT was to place large procurement orders.
As a result Europeans believed that Japanese firms might very well
develop the technologies first (they already led the world in optoelectronic components, an essential part of fiber-based broadband
systems[44]) and achieve early economies of scale in production by
supplying NTT. The Japanese threat therefore figured prominently
in European discussions of collaboration.

In short, in every branch of telematics new government programs
and regulatory developments in Japan and the United States threatened to worsen Europe's already shaky position. It became obvious
to European policy-makers, both in Brussels and in the national
capitals, that the old policies had fallen short and that new approaches would be needed for the future.

TECHNOLOGICAL CHANGE AND COLLABORATION

Patterns of Interfirm Alliances in High-Technology Sectors

Firms operating in high-technology sectors have always formed links
with other firms, frequently to obtain (or exchange) patents, manufacture a product under license, or start a joint venture. Beginning
in the late 1970s, however, the formation of these interfirm alliances accelerated. The result was an increasingly dense network of
alliances among firms, especially among those companies in the information technologies. Although there are multiple reasons for which
enterprises seek out partners, the upsurge in alliances was due in
large part to technological changes and uncertainties. As James
Thompson demonstrated, organizations attempt to stabilize sources
of uncertainty in their environment and in their core technology
(the set of techniques used to accomplish their basic productive
task).[45]

When new products and processes emerge torrentlike in a stream
of innovations (as they did in the telematics industries in the early

44. See Jonathan Joseph, "How the Japanese Became a Power in Optoelectronics," 50–51.
45. James Thompson, *Organizations in Action.*

1980s), both the task environment and the technological core of
many companies are in constant flux. It is nearly impossible to foresee what markets will develop for what products or what the competition might introduce. The dilemma is sharpened by the increasingly common phenomenon of cross-sectoral technological links.
Many products now embody innovations from fields that have
heretofore been unrelated. For instance, fiber optics (which is at the
heart of next-generation broadband telecoms systems) has brought
together companies traditionally based in electronics (like Siemens)
with firms from the glass and ceramics industry (like Corning). Thus
a company that thinks it has a stable technological core will find
itself left behind by the new combinations of technologies that are
today's hallmark. In other words, alliances are a way of dealing
with uncertainties arising from technological change.[46]

Alliances can also be a response to the needs of enterprises for
outside know-how or resources that are necessary to bring an innovation to market successfully. David Teece calls these resources
"complementary assets." Complementary assets can include technologies needed as components or inputs (technological assets) as
well as capabilities in nontechnological areas like marketing, manufacturing, and after-sales support (market-oriented assets). Gary
Pisano and Teece argue that firms can acquire both kinds of complementary assets via a range of approaches. At one extreme is the
purely market approach: purchasing the resources through "arms-
length" contracts. At the other extreme is the possibility of building
up the needed capability in-house. The purely market option runs
the risk of exploitation by the outside contractor; the in-house option is virtually certain to be impossibly expensive. Thus the middle
option: interfirm alliances.[47]

A note on interfirm cooperation is in order here. In summarizing
the growing body of research on the phenomenon, the OECD concludes that the definition of an "interfirm technical cooperation

46. Researchers at the Centre d'Etudes et des Recherches sur l'Entreprise Multinationale have constructed a large database on interfirm alliances involving European firms in a variety of sectors. They also conclude that technological uncertainties are at the heart of the recent surge in cooperative agreements. See LAREA/
CEREM, *Les stratégies d'accord des groupes de la CEE: Intégration ou éclatement
de l'espace industriel européen*, 8–15.
47. David J. Teece, "Profiting from Technological Innovation"; Gary Pisano
and David J. Teece, *Collaborative Arrangements and Global Technology Strategy:
Some Evidence from the Telecommunications Equipment Industry*, 20–30.

agreement" must be broad and inclusive.[48] Such agreements take myriad forms, beyond the well-known joint-venture structure, such as one-way and two-way patent transfers, research agreements, joint product development, licensing, and equity purchases. Mergers and acquisitions that result in the disappearance of a corporate entity do not count as cooperative agreements. At the other extreme, one-time purchases do not count. In the studies cited below, anything in between the two extremes counts. The key is that the arrangement be long-term and involve some form of collaboration.

Reasons for Interfirm Alliances in Telematics

A primary conclusion of the research on interfirm alliances in telematics is that the number of agreements involving European firms was rising dramatically in the early 1980s. The total number of such agreements per year was as follows:[49]

1980	15
1981	31
1982	58
1983	97
1984	131
1985	149

The data from the Centre d'Etudes et de Recherches sur l'Entreprise Multinationale at the Laboratoire de Recherche en Economie Appliquée (LAREA/CEREM) also show that out of a total of 587 cooperative agreements identified for the period 1980–85, 302 (or 51 percent) involved telematics.[50] As other studies prove, the phenomenon was worldwide, with American and Japanese firms joining European firms in globe-spanning, interlocking networks of alliances.[51]

48. For a discussion of definitional problems, see OECD, *Technical Co-operation Agreements between Firms: Some Initial Data and Analysis*, 6–14.
49. LAREA/CEREM, in OECD, *Technical Co-operation Agreements*, 21.
50. The LAREA/CEREM data cover four major sectors: information technologies (including telecommunications, computers, ICs, computer-aided design and manufacturing, software and services), new materials, biotechnologies, and aerospace and civil aviation. LAREA/CEREM, *Les stratégies d'accord.*
51. See Herbert I. Fusfeld and Carmela S. Haklisch, "Cooperative R&D for Competitors," 60–76; OECD, *Technical Co-operation Agreements.*

Why did telematics enterprises become so deeply involved in co-operative agreements? Researchers who have investigated the question generally agree on a set of related factors driving the phenomenon. In the discussion that follows, I distill the smallest number of separate factors possible from the broad (and frequently overlapping) lists extant.[52] The principal technological changes transforming the strategies of firms were (1) vertical technology links; (2) horizontal convergence across sectors; (3) rapid innovation; (4) escalating costs of R&D; and (5) globalization of markets.

Vertical Technology Links   Complex telematics products are increasingly designed from the components up—that is, as the number of functions that can be packed onto a chip soars because of VLSI technology, more and more of the final system can be built into the chip. Thus, in order to produce a state-of-the-art public exchange (or private branch exchange or workstation or personal computer, and so on), the manufacturer of the end product must work closely with chip designers and software experts so as to achieve the optimal balance between hardware (what is built into the circuitry) and software (what is programmed in the final system). As Borrus explains:

> Semiconductor firms now find themselves in a position where their VLSI device technology, which permits the design of logic systems in silicon, is so powerful that it forces a reconceptualizing of the design and production of final systems products. The potential impact of VLSI, in short, is to upset established design parameters in final systems, as well as in components.[53]

There were two major consequences of this trend in the early 1980s. First, the market for custom and semicustom (or ASIC) chips was booming, drawing in droves of new competitors. Over forty start-up companies entered the ASIC field in 1982–83 alone. Furthermore, the established companies in standard ICs (like Texas Instruments, Motorola, Intel, National Semiconductor, Signetics, Mostek, Harris) began to offer custom and semicustom devices.[54] Second, companies were expanding up and down the vertical chain

52. My classification of the factors behind increasing interfirm technical agreements parallels those in Sharp and Shearman, *European Technological Collaboration*, chap. 1; and Rob van Tulder and Gerd Junne, *European Multinationals in Core Technologies*, chap. 7.
53. Borrus, *Competing for Control*, 149.
54. Ibid., 152–58.

from components to final systems. Builders of computers, telecoms equipment, industrial systems, and consumer products were all acquiring semiconductor capabilities. Conversely, a number of semiconductor companies were moving into the markets for final systems—for example, Texas Instruments and National Semiconductor into computer systems and Motorola into telecommunications.[55] Frequently, the forward and backward linkages were forged through interfirm alliances. Thus, Carmela Haklisch in her study of technical agreements involving the world's forty-one largest semiconductor companies showed an increase from two such agreements in 1978 to twenty-two in 1981 to forty-two in 1984.[56]

The integrated European houses (Philips and Siemens) had always produced everything from chips to computer and telecommunications systems. But their weakness in ICs led even these giants to create links with semiconductor companies. Philips bought the American company Signetics in 1975 and struck agreements with RCA, CDC, Intel, and Siemens.[57] Siemens established ties with a plethora of firms (Table 6.5). Other linkups between systems and semiconductor companies involving European firms include Olivetti (with Zilog), GEC (Mitel), Ferranti (GTE, Nixdorf), CII-HB (Trilogy), and ICL (Fujitsu).

Horizontal Convergence across Sectors   The convergence of computers and telecommunications systems has been commonplace since the late 1970s. But the phenomenon of horizontal technology links is much broader. Microelectronics is creating overlaps among computers, telecoms, industrial automation, office equipment, and, in the near future, consumer electronics.

The convergence of data-processing and telecommunications proceeded from both ends. With the advent of digital switches, telephone exchanges increasingly resembled large computers: They processed digitized electronic impulses according to programmed instructions. Telecoms-equipment makers began to rely on technologies developed for the computer industry. On the other side, the latest development in computing, networking (or distributed processing), involved long-distance communication of data, as well as networks of computers able to share data files and programs.

55. Ibid., 165–68.
56. Carmela S. Haklisch, "Technical Alliances in the Semiconductor Industry."
57. OECD, *Semiconductor Industry*, pp. 125–37.

TABLE 6.5.   INTERFIRM ALLIANCES IN THE
SEMICONDUCTOR INDUSTRY INVOLVING SIEMENS,
1974–84

Firm	Year	Type of Link
Dickson	1974	Purchase
Intel	1976	R&D cooperation
Advanced Micro Devices	1977	20 percent stake
Litronix	1977	Purchase
Microwave Semiconductor	1979	Purchase
Databit	1979	Purchase
Threshold Technology	1980	Purchase
Intel	1982	R&D cooperation
Fuji	?	Joint venture
Philips	1982	R&D cooperation
Philips	1984	R&D cooperation

SOURCE: Organization for Economic Cooperation and Development, *The Semicon-ductor Industry: Trade Related Issues* (Paris, 1985), 125–37.

Computer networking could take the form of local area networks (linking computers directly one to another via cable) or could occur through private branch exchanges (small switches that can connect computers and other equipment like facsimile machines). In other words, data-processing increasingly relied on communications technologies.

The result of these trends was the formation of alliances between computer and telecommunications firms. For instance, the computer-industry giant, IBM, established ties for telecoms equipment (Rolm), networks (MCI and SIP of Italy), and enhanced services (with the Bundespost and NTT). The U.S. telecoms giant, AT&T, acquired a 25 percent stake in Olivetti and began marketing computers from the Italian firm.

In addition, new transmission techniques spawned other kinds of alliances for telecommunications firms. Microwave transmission brought in manufacturers of radio equipment. Satellite communications drew in aerospace companies (many with long Department of Defense experience) like Hughes, Ford Aerospace, TRW, and Lockheed. The optical-fiber transition is currently producing links between telecoms companies and makers of glass fibers and lasers. Corning, for example, has licensed its fiber technology to a number of traditional cable producers and has a joint venture with Siemens (Siecor).

The advent of programmable machine tools and automated production created opportunities for partnerships of computer companies and the makers of industrial equipment. The blending of production equipment and computers resulted in computer-integrated manufacturing (CIM): a factory in which computers control the operations of the machinery and the flow of materials through them. Agreements between automation and electronics companies include those linking Siemens to Fujitsu Fanuc, Thorn-EMI to Yaskawa, Selenia-Elsag to IBM, and Comau (a Fiat subsidiary) to DEC.[58] The next stage will likely be the transmission of high-fidelity stereo and HDTV into homes via the future broadband networks. At that point consumer-electronics companies will be working with telecoms firms and broadcasting and production interests.

The point of these examples is that as final-product markets converged and overlapped, companies were forced to expand out of their traditional activities and enter markets where they had no technical expertise. One way to acquire quickly the necessary know-how was to ally with companies already established in the field. Thus, the need to integrate technologies from a broad spectrum of sectors pushed the formation of interfirm alliances.

Rapid Innovation   Not only was technology breaking down traditional barriers between sectors, but innovation in each sector was occurring at an increasingly rapid rate. No company could predict how and when technology would change, much less cover all the necessary R&D bases. Because the basic raw material of electronics systems was the IC, rapid innovation in semiconductors meant rapid product change in all the final-use sectors. Thus, the rate of change in ICs illustrates the problem for all of telematics. In 1983 the Commission estimated that information-technology products had a life expectancy of just three years; in some subsectors, it was less than that.[59] Memory chips (specifically, DRAMs) require the great-

58. van Tulder and Junne, *European Multinationals in Core Technologies,* 240–41.

59. CEC, *Proposal for a Council Decision Adopting the First European Strategic Programme for Research and Development in Information Technologies (ESPRIT),* 51.

est density of integration and thus have driven innovation in VLSI. The 16K DRAM (16,000 bits of information) entered the market in 1976. The 64K DRAM was introduced in 1980, a quadrupling of memory capacity in four years. The 256K DRAM became available in 1982, the one-megabit chip (over one million bits) in 1984, and the four-megabit chip in 1987.[60] Thus, chip producers were able to double memory capacity approximately every two to three years.

Illustrations of rapid technological innovation abound in the telecommunications field. Public data networks (like the Minitel system in France or Bildschirmtext in Germany) did not exist in 1980, nor did services like teleconferencing and videophones. The life expectancy of products is shrinking. For instance, in the early 1980s a digital switch lasted about ten years before it was economical to replace it; the electromechanical switches replaced by the digital exchanges had a life span of thirty years.[61] The replacement rate for PABXs increased from 5 percent of total installed equipment per year to between 10 and 20 percent in the early 1980s.[62] And now progress in optoelectronics is leading to a whole new generation of switches and terminals based on photons instead of electrons. Makers are under severe pressure to provide their customers with upgrades of technical quality comparable to those of competing producers and to provide them as rapidly.

In this environment in the early 1980s no firm could hope to stay abreast of all the market-making developments by itself. Partnering was a way of reducing risks, as one partner might pick up on trends that the other partner missed. Or, in other words, in an era of rapid technological innovation, interfirm alliances reduced the risk of missing out on a crucial development.

Escalating Costs of R&D   Given the spectrum of technologies involved and the rapid pace of change, it became increasingly difficult for any one firm to muster the financial and human resources needed to develop new generations of products. For instance, as chip complexity went up and the line width on ICs went down (to below one micron), the development cost of a new chip soared. The one-kilobit memory chip cost about $2 million to develop; the one-

60. See Borrus, *Competing for Control*, 176.
61. A. D. Little, *European Telecommunications*, 39.
62. OECD, *Telecommunications*, 77.

megabit RAM cost around $100 million. Whereas development costs for the four-bit microprocessor ran about $15 million, the most recent models cost about $150 million. Furthermore, increasingly complex logic chips with one million or more elements are impossible to design without specialized computer programs (computer-aided design, or CAD).[63]

The need for such programs raises the problem of software. Software (or the programmed instructions needed to run all telematics systems) was becoming the costliest part of R&D and final products by 1983. While hardware costs (on a per-function basis) had declined steadily, software costs (on a per-line basis) had remained constant or even increased. Discounting for inflation, a line of programming in the early 1980s cost about $10 to $50, about the same as it had in 1955. The problem was that current systems employed more lines than before, sometimes hundreds of thousands of lines. Thus, computer firms devoted more than half their R&D resources (money and manpower) to software, and software accounted for well over half the cost to the user of a final computer system.[64] The Commission estimated that although software accounted for 20 percent of the cost of R&D for a public exchange in 1970, it would reach 80 percent of the R&D cost by 1990.[65] In the early 1980s software already accounted for over 60 percent of R&D on public exchanges.[66]

Finally, telecommunications equipment illustrates the soaring costs of R&D in final systems. ITT spent $30–$40 million to develop its Pentaconta switching system in the early 1960s; it spent $300–$500 million on its 1240 system in the late 1970s.[67] By 1986 the R&D expense associated with developing a current-generation digital public exchange ranged from $500 million (Ericsson's AXE) to $1.4 billion (GEC/Plessey's System X).[68]

Rising R&D costs motivated companies to seek partners that could share the burden. The constraints on R&D resources were not always financial either. In Europe the primary bottleneck might well have been in the supply of qualified scientists, engineers, and tech-

63. See OTA, *International Competitiveness in Electronics*, 77.
64. Ibid., 86–87.
65. CEC, *Towards a Dynamic European Economy*, 90.
66. OECD, *Telecommunications*, 54.
67. Ibid., 54.
68. Godefroy Dang Nguyen, "Telecommunications: A Challenge to the Old Order," 108.

nicians. As one executive of a German computer company told me, in Europe the scarcest resource was trained personnel, and that scarcity motivated many of the alliance strategies of European firms.[69]

Globalization of Markets   Because of high R&D costs, companies had to have vast sales to amortize the investment in R&D. Thus, telematics markets were becoming increasingly global. Alliances were the means of getting inside a market protected by official and unofficial barriers to trade. A partner established in a national market could sell the goods of an outside company either directly or as a second source. Some firms sought allies for access to their distribution networks. In their study of the telecommunications-equipment sector, Teece, Pisano, and Michael Russo divided the motivations for seeking interfirm alliances into two main categories: technology access and market access. Their data (from Futuro Organizzazione Risorse in Rome) showed that out of 117 total agreements, distribution/marketing was the primary motive in 35 (29.9 percent). Distribution/marketing was joined with other objectives (R&D, production) in 15 more agreements, meaning that the market-access motive applied in 42.7 percent of the agreements. Technology access was the sole motive in 36 agreements (30.8 percent) and figured in 47.0 percent of the agreements overall.[70]

LAREA/CEREM data showed an even greater percentage of agreements in the overall telematics sector having marketing considerations as their primary motive. Marketing was a factor in 39 percent of 316 agreements, while knowledge generation (technology) figured in 19 percent and production in 16 percent.[71] It should be noted that marketing motives frequently tie directly to technology concerns, as in the marketing of another firm's product in order to fill out a company's product range. Again, it is increasingly difficult for a single enterprise to cover all the technology bases it needs.

As a result of these five factors, alliances linking American, Japanese, and European companies multiplied during the 1980s. Some studies showed that European firms chose American partners more readily than Japanese or European partners, though that gap was

69. Interview 46.
70. David J. Teece, Gary Pisano, and Michael Russo, *Joint Ventures and Collaborative Arrangements in the Telecommunications Equipment Industry*, 27, 63.
71. LAREA/CEREM, *Les stratégies d'accord,* 47.

TABLE 6.6.   NUMBER OF INTERFIRM ALLIANCES IN
ELECTRONICS THROUGH 1984

	Before 1982 (Cumulative)	1982	1983	1984
Europe	22	7	23	33
Europe-United States	61	34	36	43
Europe-Japan	17	12	5	5

SOURCE: Réseau (Milan), in Organization for Economic Cooperation and Development, *Technical Cooperation Agreements between Firms: Some Initial Data and Analysis* (Paris, 1986), 19.

beginning to close for intra-European alliances (see Table 6.6). Explanations for the evident early preference on the part of European firms for American rather than other European partners abound.[72] Some of them are sociocultural: European firms were too accustomed to seeing each other as competitors to think about cooperating. Others are more technological: American firms frequently possessed the most advanced technologies. Still others stress market access: Links with American firms provided a way to enter the huge American market. What is important for this study is that around 1980 few interfirm accords linked European companies. Whatever potential existed for such ties had not been explored, much less exhausted.

SUMMARY

This chapter has detailed the origins of the shift in Europe from national-champion strategies toward collaboration. Two principal factors drove the movement toward collaboration. First, intense international competition in telematics fueled a crisis in European telematics policy-making. European enterprises continued to fare poorly in competition with United States and Japanese firms. After a de-

72. What is not so clear is why there was not a boom in alliances between European and Japanese firms, especially given the technological strengths of the Japanese. Answering that question would require research into the perceptions and decision-making of European corporate leaders, a task beyond the scope of this study. It may be that Japanese firms were not as eager to ally or were less willing on average to grant access to their technologies.

cade of national-champion policies Europe's share of its own markets was declining in semiconductors and computers. Even the traditional European stronghold, telecommunications, was slipping. Contrasting with Europe's failures, Japan had risen from behind Europe to technological prominence. Finally, new government-led R&D programs in the United States and Japan and the freeing of IBM and AT&T to compete in new markets threatened to sink Europe even deeper in its hole.

Second, technological changes also disposed telematics companies to seek interfirm alliances. European enterprises around 1980 were becoming heavily involved in strategic alliances, especially with American companies. Technological change assumes critical importance in the argument I develop here: Because technological changes motivated European companies to seek interfirm alliances, the major European telematics enterprises were receptive to Commission initiatives in support of collaboration. In an important sense, therefore, technological change paved the way for the formation of the transnational industrial coalition that was the heart of ESPRIT and RACE.

SEVEN

ESPRIT

Opening the Door

By 1980 the dimensions of Europe's crisis in telematics were becoming clear. A decade and a half of national-champion strategies had failed to close any gaps between Europe and the telematics leaders, Japan and the United States. Europe faced the prospect of a neck-and-neck race between American and Japanese industries that would leave Europe eating dust. This was the challenge facing Europe's policy-makers at the beginning of the new decade.

My argument is that the crisis led European elites to question their unilateral, national strategies for promoting telematics industries. Policy-makers in each of the major countries entered what I have labeled the adaptive mode: They were all searching for new approaches. The evidence consists of high-level, high-profile, official studies of telematics policies in the United Kingdom, France, and the Federal Republic of Germany. In each case blue-ribbon commissions produced new telematics plans. Simultaneously, the same governments were agreeing to collaborate in the Commission's ESPRIT program. The evidence is clear: Across Europe, governments were in a state of crisis-induced adaptation in the early 1980s.

The previous chapter also spelled out the technological changes that underpinned the intensified interest of private enterprises in interfirm alliances. In this chapter, I show how a transnational coalition of the major telematics companies combined with an entrepreneurial Commission of the European Communities to produce a collaborative response to the crisis. To be sure, collaboration was by no means the necessary outcome of the policy adaptation going on in the various countries. Collaboration emerged because the Commission and the industry coalition (the Twelve Roundtable companies) were able to gain the necessary political support from governments looking for new directions in telematics policy.

CRISIS-INDUCED ADAPTATION

Although collaboration eventually emerged in the form of ESPRIT, it did not replace national programs to build up domestic IT industries. In fact, national programs to support IT flourished with renewed vigor in the early 1980s. In this first section I describe the adaptive process in France, Germany, and the United Kingdom, focusing on changes in national policies. It bears remembering that although each country adopted new national strategies, the ESPRIT program was being created during the same period (1982–1984). In other words, collaboration did not replace national policies but took its place beside them.

Several themes link the national cases. First, each government believed more firmly than ever in the importance of IT in driving economic development and growth. Even those governments ideologically opposed to public intervention in industry (the Helmut Kohl government in Germany and the Thatcher government in Great Britain) provided massive new supports to the IT sectors in the early 1980s. Second, government officials, company executives, and Brussels technocrats all spoke in the early 1980s of the urgent necessity of meeting the Japanese and American challenge.

Third, as Malerba points out, government and industry had come around to the view that strength in telematics required a presence in the design and manufacturing of advanced semiconductors. Both business and government had missed the boat the first time around, and Europe by the late 1970s was absent from world microprocessor and memory-chip markets. Only then did governments conclude that "a domestic productive capability in LSI devices was of strategic value for reasons of national security and for *sustaining the electronic and the manufacturing industries as a whole.*"[1] Although it is not his analytic concern, Malerba thus provides evi-

1. Malerba, *Semiconductor Business,* 162, my emphasis.

dence of the kind of policy adaptation I expect. Further confirmation of this adaptation comes from a well-placed observer, Pasquale Pistorio, president of the Italian firm SGS: "European governments are beginning to see that they've got to protect this very important industry [semiconductors] if they're going to preserve the quality of the European electronics industry."[2] The programs described in this section show that governments were looking for new ways to support their domestic firms' reentry into advanced digital microelectronics.

The semiconductor companies themselves were also trying to return to advanced digital ICs; by the early 1980s Siemens, Philips, Thomson, SGS, and Inmos were all starting again to produce highly integrated memory devices.[3] A *Financial Times* survey of the European electronics industry in 1984 summarized the by-then general consensus that the production of standard ICs (especially memories) was essential as it drove product, design, and production innovations.[4] Furthermore, as final electronics systems were increasingly built into the chips, the competitiveness of European systems makers increasingly depended on mastery of chip technologies. Executives from Europe's major semiconductor firms confirmed this observation. Klaus Ziegler, vice-president of the Components Group at Siemens, explained his company's efforts starting in the early 1980s to rejoin the competition in advanced semiconductors: "First, for a company like us that integrates microelectronics into its finished products, it's essential to avoid becoming dependent on our competitors. Secondly, increasing system know-how is going into components all the time, and we're justifiably reluctant to have our expertise siphoned off by others."[5] Cees Krijgsman, managing director for ICs at Philips, speaking of Europe's position in electronics and advanced manufacturing more generally, declared, "To be competitive, we simply have to be strong in semiconductors."[6]

Fourth, and last, promoting collaboration among companies and between industry and academia became a feature of European IT

2. Quoted in Thane Peterson, with Amy Borrus, "Europe's Chipmakers Pull Out of a Long Losing Streak," 22.
3. Paul Betts, "Thomson Has Another Try," *Financial Times,* 25 October 1985, p. 18.
4. Guy de Jonquieres, "Electronics in Europe," *Financial Times,* 28 March 1984, Survey, p. 1.
5. Siemens, "Japan's Challenge—Europe's Response: Interview with Klaus Ziegler," 18.
6. Quoted in Peterson, "Europe's Chipmakers," 22.

support programs. Such links were a central part not only of ES-
PRIT but also of national programs in Germany and the United
Kingdom. Collaborative R&D was perceived as a crucial part of
American and especially Japanese successes in promoting high-technology industries. In a sense, then, collaborative R&D in Europe
was an adaptive response based on imitation of the acknowledged
leaders. Table 7.1 presents a chronology of the major events in IT
policy.

France

French electronics and data-processing programs in the first years
of the Mitterrand government were even more nationalist than previous ones, difficult as that may be to imagine. The French trade
deficit in electronics goods doubled from 1981 to 1982, to FFr 12
billion.[7] President Mitterrand was determined to reverse that trend.
The Giscard government, prior to leaving office, had commissioned
a series of studies on French high-tech capacities. The new government expanded these studies and used them as a basis for the *colloque national* on research in January 1982, designed as a forum
in which those interested (unions, industries, banks, researchers, civil
servants) could provide input for a new science and technology policy. The new policy emerged in July 1982 with the *Loi d'orientation
et programmation* (LOP). The LOP laid out ambitious goals to make
France "the world's third scientific power" by the 1990s.[8] The overall
goal was to spend 2.5 percent of GNP on domestic R&D by 1985,
from a base of 2.01 percent in 1981, with the effort focused on
seven high-priority *programmes mobilisateurs*.[9] One of the priority
programmes was for the electronics sector, the *filière électronique*.
    Simultaneous with the LOP process the government set up within
the Ministry of Research and Technology a special *Mission filière
électronique* to study that sector in particular. Under Abel Farnoux
(who lent his name to the report) the *Mission* submitted its conclusions in March 1982. The Farnoux Report declared, "If France
is to maintain its independence, master the new communications
systems and put the crisis behind it, it is imperative for the country

    7. David Marsh, "Bid to Narrow the Gap," *Financial Times,* 21 March 1983,
Survey, p. 6.
    8. OECD, *Innovation Policy,* 67.
    9. Ibid., 71–75.

TABLE 7.1.   ESPRIT CHRONOLOGY

Year	Event
1979	EEC: In November the Commission's IT Task Force proposes a European telematics strategy.
1980	EEC: In February Davignon meets with telematics-industry executives.
	EEC: Study on Europe's needs in long lead-time R&D in information technology begins.
1981	United Kingdom: Thatcher government appoints Kenneth Baker minister of state for information technology.
	EEC: In November the Research Council approves the Commission's Microelectronics Program.
	EEC: Davignon invites directors of the twelve largest telematics firms to Roundtable discussions.
1982	United Kingdom: 1982 is declared the "Year of Information Technology."
	France: In March the Farnoux Report lays the basis for revamping telematics policy.
	EEC: In May the Commission proposes the ESPRIT program.
	EEC: In August the Commission proposes the ESPRIT pilot phase.
	EEC: In December the ESPRIT pilot phase is formally approved.
1983	United Kingdom: In the spring the Alvey Report proposes what will become the Alvey Program.
	EEC: In June the Commission proposes ESPRIT Phase I.
	France: In September President Mitterrand circulates a memorandum advocating a "European industrial space."
1984	Germany: BMFT announces a new IT program; Mega project is launched (Siemens and Philips).
	EEC: In February ESPRIT Phase I is approved by the Council.
1986	EEC: All funds for ESPRIT Phase I are allocated.
	EEC: In May the Commission proposes Phase II.
1987	EEC: The ITTF becomes DG XIII.
	EEC: In July the Council approves in principle the new Framework Programme for R&D after a one-year delay.
1988	EEC: In April the Council formally approves ESPRIT II.

to master the key activities of the electronics sector."[10] The report became the basis for the *Programme d'action pour la filière électronique* (PAFE). The PAFE envisioned a wholesale revamping of French telematics policy, with the different sectors seen as parts of a single whole extending from ICs to final systems in communications, consumer electronics, data-processing, and office and factory automation. Again, ambitious national goals were the order of the day. For the period 1983–87 France would shoot for a production increase of 9 percent (versus the actual trend of a 3 percent increase), 50,000 new jobs in the sector (trend: 10,000 lost), and a trade surplus of FFr 14 billion (trend: FFr 20 billion deficit by 1987).[11] This program would be accomplished via several major actions:

1. Restructuring through nationalization

2. Increased R&D spending

3. Diffusion of technologies through large national projects

4. Obligating French firms to buy French components and equipment where possible[12]

5. Restructuring the education system to produce additional skilled personnel

Structuring the program through national projects resembled the Japanese model for mobilizing whole industries.[13]

In semiconductors French goals were ambitious: The microelectronics segment of the *filière électronique* aimed at French independence in semiconductor technology by 1986. The government was convinced that "without integrated circuits, the entire French electronics industry would crumble."[14]

The report also mentioned the possibility of eventually including firms from other European countries in the national projects and

10. Quoted in ibid., 218.
11. Ibid., 219.
12. Such an obligation was apparently quite common in the early 1980s in France. Users of both ICs and computers were pressured to buy French (Kenneth Dreyfack, "France Wants Bigger Piece of Pie," 98; John Morris, "France Plans DP Sell-Off," 66).
13. Robert T. Gallagher, "French Want National as Partner," 104.
14. Quoted in Thomas R. Howell et al., *The Microelectronics Race*, 170.

possibly even American or Japanese firms in the long term.[15] But, as the OECD study of innovation policy in France points out, the "PAFE was not set in an international context from the outset. It was not until over a year later after the programme was adopted in September 1983 that France made the first attempts at achieving some complementarity between the French vision of the electronics sector and European action."[16] Indeed, President Mitterrand announced early in his term that his objective was "reconquering the domestic market" and making France the third IT power behind the United States and Japan.[17]

The government reiterated in September 1983 its intention to invest FFr 140 billion over five years to achieve what Mitterrand called the "great electronic leap forward." Minister for Industry and Research Laurent Fabius declared, "Meeting the challenge of electronics and information technologies is the number one priority in industrial policy for the country."[18] The total investment included FFr 60 billion from the state, the rest to come from industry.[19] The state contribution to R&D support was to total FFr 30 billion, approximately double what it would have contributed otherwise. The state contributions would come from several ministries (largest shares from defense and the DGT) and take the form of contracts, grants, and equity funding. In practice the funding channels were extremely complex, so it was hard to tell whether the spending targets were attained.[20] They almost certainly were not.

France's inability to meet its targets was behind its sudden about-face in 1983 and 1984. When it became clear that France could not possibly afford such an ambitious and nationalist strategy, the French government became a vigorous advocate of European collaboration. How this turnabout occurred is instructive, for it demonstrates the recognition on the part of a government of the failure of a national-champion approach and the subsequent turn toward coop-

15. Robert T. Gallagher, "France Urged to Reorganize Electronics," 105.
16. OECD, *Innovation Policy,* 219.
17. David Marsh, "French Computer Strategy Under Fire," *Financial Times,* 15 April 1983, p. 32; and David Marsh, "Stronger Emphasis on Joint Ventures," *Financial Times,* 11 July 1984, Survey, p. 7.
18. "M. Mitterrand: la France proposera un plan européen de haute technologie en 1984," *Le Monde,* 27 September 1983, p. 44.
19. Paul Betts, "France Campaigns for Electronics Collaboration," *Financial Times,* 28 September 1983, p. 1.
20. OECD, *Innovation Policy,* 221.

eration. What is surprising is how quickly the shift occurred: The
Farnoux Report was issued in March 1982; by September 1983 the
French government was beginning to speak out for European high-
technology cooperation.

As the OECD report on France points out, economic growth rates
were far lower in the early 1980s than the government had predicted. The economy was not producing at the level needed to support the huge investments planned for the *filière électronique.* Evidence of trouble came in July 1983, when a large chunk of the
PAFE program was placed under the DGT. This switch meant that
DGT revenues from its profitable network activities (which were
independent of the national budget) would be used to underwrite
the electronics plan, including huge subsidies to loss-making firms
like Bull and Thomson and in addition to the usual funding for
telecoms suppliers like CGE and CGCT. Not surprisingly, the re-
shuffling quickly ruined the DGT's former extremely strong financial position.[21] The following year the Direction des Industries Elec-
troniques et de l'Informatique, which had been in charge of the
filière électronique, was quoted as saying, "Unfortunately, in the
present context, we cannot afford to make the effort required. Certainly for the new [microelectronics] plan which will start in 1987,
a European effort will have to be considered."[22] In short, France
was rapidly discovering the limits to what one European nation,
even one of the largest, could hope to achieve autonomously in high
technology.

Thus, by the fall of 1983 the French were already beginning to
sing a cooperative tune. The Mitterrand government in September
1983 circulated a memorandum among its fellow EC members advocating *"un espace industriel européen."* President Mitterrand announced at the end of the month that France was preparing to propose a European plan for high technology, especially IT. The French
president declared in a speech: "We should show the same ambition
in high technology; our independence as Europeans and our iden-
tity are the stakes. Otherwise we will be submerged by competitors
like the United States and Japan."[23]

21. Ibrahim Warde, "French Telecommunications," 109–10.
22. Howell et al., *Microelectronics Race,* 171.
23. "M. Mitterrand: la France proposera un plan européen de haute technologie
en 1984," *Le Monde,* 27 September 1983, p. 44.

By this time France also had a new minister for industry and research, Fabius, who was a strong advocate of increased European R&D collaboration. Fabius had replaced Jean-Pierre Chevènement, who was in office during the preparation of the LOP and the PAFE and who was a vigorous proponent of French technological independence. Indeed, Chevènement had issued instructions that French scientists publish their research in French and defend French as a scientific language in international meetings.[24] Chevènement resigned amid controversy over his interventionist approach to the nationalized industries in March 1983 and was replaced by Fabius.

As Mitterrand began talking up European cooperation on high technologies, Fabius also took up the theme. The minister for industry and research spoke out in favor of European collaboration in electronics and telecommunications, including the gradual opening of public procurement markets. The government also expressed the view that alliances among European industrial groups should be favored over those with non-European firms.[25] The French even criticized an AT&T-Philips deal as a Trojan horse for the American company.[26] At the same time, however, French electronics firms were striking alliances with U.S. companies—Thomson with National Semiconductor, Matra with Harris. Nevertheless, from that point on the French became consistent supporters of the Commission's efforts leading up to ESPRIT.[27]

Just over a year later, Fabius ascended to the post of prime minister. Mitterrand's selection of Fabius reflected both his hopes for the technological modernization of French industry and his growing commitment to European cooperation. As chairman of the Council of Research Ministers of the EC during the first half of 1984 Fabius had been a staunch supporter of the Commission's technology plans and had been a consistent backer of ESPRIT, whose first phase was approved during Fabius's tenure. To fill his former cabinet position, Fabius chose Hubert Curien, though the industry brief was moved

24. John Walsh, "France Readies New Research Law," 712.
25. Paul Betts, "France Campaigns for Electronics Collaboration," *Financial Times*, 28 September 1983, p. 1.
26. Guy de Jonquieres, "Government Aims to Float Off Slice of BT in U.S.," *Financial Times*, 20 September 1983, p. 6.
27. Paul Betts, "Government Encourages Closer European Collaboration," *Financial Times*, 11 July 1984, Survey, p. 2; and David Marsh, "Stronger Emphasis on Joint Ventures," *FT*, 11 July 1984, Survey, p. 7.

elsewhere and the ministry once again covered research and technology. The appointment of Curien was also significant. Curien was at the time president of the European Science Foundation and was a former chairman of the council of the ESA. In other words, Curien was a committed and experienced Europeanist on technology matters.[28] Thus, two key positions in the Mitterrand cabinet after July 1984 were filled by advocates of European collaboration. Curien would later be one of Mitterrand's point men in selling EUREKA to France's partners. In barely three years the French government had evolved from a vigorous promoter of national champions and independence into an enthusiastic advocate of European collaboration.[29]

Federal Republic of Germany

Germany, like the United Kingdom, frequently finds its desire to advance its IT industries at odds with its noninterventionist instincts. The Ministry of Economics tends to uphold market solutions to industrial problems, while the Ministry of Research and Technology (BMFT) has been willing to formulate programs to stimulate high-tech sectors. The CDU/CSU-FDP coalition that has governed since 1982 is ideologically uncomfortable with state intervention. Still, the Kohl government, like its SDP predecessors, has instituted programs to stimulate the IT industry.

   In the early 1980s the BMFT's Microelectronics Program, begun in 1979, was supporting R&D for chip design, production technology, materials, and applications. Financial support shifted emphasis from data-processing equipment to microelectronics and applications. In 1980 state funding for the computer sector was more than twice that for microelectronics; in 1983 the ratio was 5:2 in favor of microelectronics.[30] In addition the government agreed in 1984 to help underwrite the Mega project involving Siemens and Philips, whose object was the production of one-megabit static random-access memories (SRAMs) and four-megabit DRAMs. The German government would contribute DM 300 million (Siemens

28. David Dickson, "France's New Technocrats," 486–87.
29. See Guy de Jonquieres and Paul Betts, "The Euphoria Is Over," *Financial Times,* 7 February 1985, p. 14.
30. Erik Arnold and Ken Guy, *Parallel Convergence: National Strategies in Information Technology,* 142.

claimed to have allocated DM 2.2 billion for the project, including investment in a new plant).[31] The German government saw micro-electronics as one sector whose fate could not be left entirely to the operations of the market.

The BMFT announced in 1984 a new *Informationstechnik* program.[32] The plan had been long awaited, held up in part by objections from the Ministry of Economics.[33] The new program allotted DM 3 billion ($1.14 billion in 1984) over five years, from 1984 to 1988. The government declared that its IT program was a necessary response to technological competition from the United States and Japan.[34] The minister for research and technology, Heinz Riesen-huber, declared:

> Our share of the world market in high technology goods is stagnating, while Japan's share in the last decade has doubled. We cannot overlook this, if we want to maintain our leading position in the 1990s. . . . To this end, the government has approved the comprehensive plan to support microelectronics, information and communications technology. . . . In this plan, we are documenting our determination to accept the challenges of information technology and to improve our competitiveness.[35]

The BMFT report noted that IT production (components, consumer electronics, communications, computers, industrial automation) was growing far more rapidly in the United States and Japan than in Germany. The position of the German semiconductor industry was unsatisfactory: Whereas the United States and Japan were exporting ICs to the world, German industry supplied only 60 percent of the country's demand for ICs.[36]

The *Informationstechnik* program was the first German IT support scheme in which industry was consulted in the planning stages; fifteen companies, including both large and small ones, helped prepare the program.[37] Though the plan specified assistance only for

31. Guy de Jonquieres, "Philips Joins Siemens in Microchip Project," *Financial Times,* 11 October 1984, p. 1.

32. This announcement came at about the time the first year of the full ESPRIT program was getting underway.

33. Arnold and Guy, *Parallel Convergence,* 140.

34. Jonathan Carr, "Bonn Wakes Up to the Challenge of the Chip," *Financial Times,* 28 March 1984, Survey, p. 6; Interview 4.

35. Quoted in Howell et al., *Microelectronics Race,* 174.

36. Rupert Cornwell, "Bonn Bid to Push Ahead in High-Technology Field," *Financial Times,* 15 March 1984, p. 3.

37. John Gosch, "Germany Spends for Its High-Tech Future," 101.

R&D, and only up to 50 percent of project costs, the government indicated that state procurement, both civil and military, would provide a further stimulus to industry. The Bundespost alone was to spend a predicted $750 million annually on broadband telecommunications. The largest chunk of the DM 3 billion would go for components R&D (DM 1.41 billion), including a DM 608 million project on submicron-chip technologies. Other major areas were data-processing, industrial automation (CAD and CAM), and telecommunications.[38] A final noteworthy aspect of the program was that it would emphasize cooperation among companies and between industry, universities, and research institutes. Indeed, collaboration was becoming a keyword in German technology policy.[39] Two executives from major German electronics firms told me that interfirm collaboration had been stressed by the BMFT in its programs since 1982.[40]

However, nothing like the enthusiasm of the French for European collaboration evolved in Germany. Rather, the government was divided internally. Foreign Minister Hans-Dietrich Genscher (of the FDP) emerged as the most avid and consistent proponent of European collaboration. As one official in the BMFT put it, "Since 1984, Genscher has mentioned technology in nearly every speech. He always speaks of the necessity of European cooperation in space and information technologies." This official noted a high level of concern for technology questions ever since Genscher appointed Konrad Seitz as director of his planning staff at the Foreign Ministry.[41] The Finance Ministry, in contrast, consistently questioned the value of European collaborative programs (including the EC's Framework Programme and EUREKA, for example) and sought to restrict their budgets. The BMFT favored collaboration for basic research, as in fusion, space, and synchrotron radiation. For more commercially relevant technologies, the BMFT argued for selectivity, so that European programs would complement national ones. Thus, the BMFT saw the major components of the Framework Programme (including ESPRIT) as essential but would have preferred to reduce the budget for other programs.[42]

38. Ibid., 101–2.
39. Rupert Cornwell, "Bonn to Spend DM3bn on High Technology Boost," *Financial Times*, 9 March 1984, p. 1; Arnold and Guy, *Parallel Convergence*, 141.
40. Interviews 45 and 46.
41. Interview 4.
42. Ibid.

United Kingdom

Great Britain during the Thatcher years had been torn between an anti-interventionist ideology and an increasing perceived need to support the IT industries. These contrary impulses led to what Erik Arnold and Ken Guy call "stop-go policies."[43] Thus, the Thatcher government began with measures to privatize that part of the industry that had come under state tutelage. The National Enterprise Board was transformed into the British Technology Group and told to sell off its shares in ICL and Inmos. The government announced that a £35 million cash infusion for Inmos in January 1984 would be the last.[44] After offers for Inmos from AT&T and a Dutch consortium had been turned down in 1984, Thorn-EMI bid £144 million, raised later to £192 million, for the state's 75 percent interest in the company. This proposal was approved. The same month (August 1984) the electronics and telecommunications conglomerate STC purchased ICL for £653 million; ICL had required government loans and loan guarantees in 1980–81, though it had achieved a profit in 1981–82. Some political opposition arose to the deal because 35 percent of STC was owned by the American firm ITT; but the government approved the acquisition after ITT announced a reduction of its interest to 25 percent.[45]

In late 1980, however, the government asked the ACARD to report on whether the state should promote the IT industries in the United Kingdom. The government decided that it should, and in early 1981 Thatcher appointed a minister of state for information technology, Kenneth Baker. He immediately announced a "Year of Information Technology" in 1982. The DTI began that year to fund R&D through its Support for Innovation program (SFI). The importance attached by the DTI to IT is witnessed by the large share of its SFI spending that went to IT: In 1983–84, out of £322.53 million, £206.4 million supported IT projects.[46] However, after a change of minister in September 1984, government support for IT moved from funding specific projects to more indirect activities. The

43. Arnold and Guy, *Parallel Convergence,* 120.
44. In all, Inmos had received £211 million in government funds, about twice the amount initially foreseen, and had not turned a profit. Still, the company was generally regarded as technologically advanced if limited by a restricted product range (Kevin Smith, "Inmos Forced to Get Off the Dole," 106).
45. Hart, "British Industrial Policy," 154.
46. Ibid., 121.

total DTI budget for industrial R&D in 1985–86 was £176.6 million, of which £97.7 million was earmarked for IT.[47]

Britain's major new support plan for IT in the 1980s was a reaction to Japan's 5G program.[48] British delegates in attendance at the conference in which Japan made its announcement returned gravely pessimistic about British prospects in the face of such an effort. The government quickly ordered a study of the United Kingdom's IT industries, to be carried out by an independent committee chaired by John Alvey, technical director of British Telecom. Reporting within months, the committee proposed a major program of cooperative R&D to be sponsored by the government. In addition the R&D would focus on areas that were of industrial interest but still some distance from production for the market; this vague category of research became known, especially within the ESPRIT program, as precompetitive.

Whitehall fretted about industrial intervention but in the end approved the Alvey Program after cutting back some of the recommendations. The Alvey Committee had requested 60 percent funding for industrial R&D; the government reduced the level to 50 percent, apparently a personal decision by Thatcher.[49] The government also slimmed the proposed Alvey Directorate from thirty persons to about six, and the total government contribution from about £250 million to £200 million.[50] The funds came from the Ministry of Defense (£40 million), the Department of Education (£50 million via the Science and Engineering Research Council), and the DTI (£110 million).[51] Industry would contribute £150 million, bringing the total budget to £350 million (about $550 million at the time). This budget dwarfed that of any previous civil R&D program. As Patrick Jenkin, secretary of state for industry, told the House of Commons, "This is the first time in our history that we shall be embarking on a collaborative research project on anything like this scale."[52]

47. Ibid., 123.

48. See Freeman, *Technology Policy*, 126.

49. Alan Cane, "Britain Enters the Great Race," *Financial Times*, 9 May 1983, p. 12.

50. "UK Fifth-Generation Computer Program Passes Final Hurdle," 76; Jason Crisp, "State Joins Industry in £350m Research on Super Computers," *Financial Times*, 29 April 1983, p. 1.

51. Jason Crisp, "State Joins Industry . . . ," *Financial Times*, 29 April 1983, p. 1.

52. Alan Cane, "Britain Enters the Great Race," *Financial Times*, 9 May 1983, p. 12.

Alvey followed the Japanese model, being designed to encourage collaborative research among U.K. companies, universities, and laboratories.[53] As Jenkin observed, "Industry, academic researchers, and government will be coming together to achieve major advances which none could achieve on their own."[54]

The technical aims of Alvey fell into four major areas: VLSI technology and design tools, software engineering, "intelligent knowledge-based systems" (AI), and person/machine interfaces (display technology, speech- and image-processing). Note again the prominence of advanced microelectronics: Alvey had asserted that the United Kingdom needed access to world state-of-the-art VLSI technology.[55]

The British Alvey planners were able to coordinate the program with ESPRIT, which was emerging at the same time, so as to avoid excessive overlap between the technical objectives of the two programs. Such planning was possible in large part because the head of the Alvey Directorate, Brian Oakley, was a U.K. member of the ESPRIT Management Committee at the earliest stages.[56]

Other Countries

The big three were not the only European countries with a growing interest in IT. Italy in the mid-1980s began a program for R&D in VLSI technologies that focused on developing prototype ICs, CAD for chips, and process technologies. SGS, the major Italian semiconductor firm (before its merger with the commercial semiconductor activities of Thomson), was slated to receive 13 billion lire in subsidies and 13 billion lire in soft loans.[57] The Dutch government agreed to chip in DM 200 million (compared with West Germany's DM 300 million) for the Mega project involving Philips and Siemens; Philips took charge of the part of the project aiming at producing one-megabit SRAMs.

Even Belgium was getting into the act. The Belgians began construction of a state-of-the-art microelectronics research laboratory at Katholieke Universiteit Leuven in 1984. Called IMEC for Interuniversity MicroElectronic Center, the facility was designed to concentrate and optimize the electronics research going on at Belgium's

53. Kevin Smith, "UK Pursues Fifth-Generation Computer," 101.
54. Ibid.
55. Ibid., 102.
56. Interview 47.
57. Howell et al., *Microelectronics Race,* 175.

three Flemish-speaking universities. IMEC's research would cover advanced processing, packaging, VLSI design methods, and opto-electronics. Sixty percent of the funding would come from the Belgian government and 40 percent from industrial research contracts. The government already had in place a program to train personnel from Belgian companies in advanced microelectronics.[58] Two years later Belgium's first native IC maker came into being. The new enterprise, Mietec, was a joint venture between a development-capital company owned by Flanders state and Bell Telephone Manufacturing, ITT's Belgian subsidiary. The national government did not participate, as it had in IMEC. Mietec was given the best available fabrication equipment and the capacity to increase production rapidly.[59]

To conclude this section, despite national differences of approach and emphasis, common strands run through each of the national IT support programs of the early 1980s. First, in the early 1980s policy-makers in France, Germany, and the United Kingdom all established high-profile commissions to search for new IT policies (Alvey in the United Kingdom, Farnoux in France, the BMFT report in Germany). I take this as evidence that policy-makers were responding to the perceived failure of previous (national-champion) policies by searching for new approaches; in other words, policy-makers had entered the adaptive mode. Second, even those states ideologically indisposed to intervene (Britain, Germany) did so and at unprecedented levels. I take this as evidence of the importance European governments attached to the IT sector, especially to advanced microelectronics (VLSI).

Third, in each case the national program was a response to an acute (Japanese and American) threat to national capabilities in the IT sectors. Fourth, promoting collaboration among industry, universities, and research laboratories was a feature of national programs, most notably in Britain's Alvey but also in Germany and Italy. Finally, at the same time they were devising new nationalist responses to the high-tech challenge, each country also agreed to unprecedented degrees of international collaboration in IT. France became the most vigorous supporter among the big three of Eu-

58. Robert T. Gallagher, "Belgians Set Up Advanced IC Lab," 18–19.
59. Robert T. Gallagher, "Now Belgium Has Its Own Player in the IC Game," 20–22.

ropean IT collaboration. Even the reticent governments eventually agreed: Europe faced a crisis in the information technologies that were crucial to continued economic growth, and European collaboration could be one dimension of the response.

THE GENESIS OF ESPRIT

Officials within the Commission were by the second half of the 1970s already convinced that Europe had to cooperate in IT in order not to lag ever further in that economically crucial area. But national governments during that decade were completely unreceptive to Commission appeals. At the end of the 1970s the mounting disillusionment with past policies coincided with the appointment of a new commissioner of industry in Brussels, Davignon. Davignon was the entrepreneurial IO official par excellence. He was able to take the arguments for cooperation developed within the Commission and sell them first to industry and then, with the help of the powerful business coalition, to national governments. Davignon was energetic, passionately committed to European collaboration in telematics, and politically savvy. But I do not want to overstress his role, important as it was.

The Commission initiative for ESPRIT bore fruit because of several factors: Purely national policies had failed; national policymakers were in adaptive mode; the major industrial actors were for their own reasons interested in cooperation; and an activist Commission mobilized a political coalition behind collaboration.

ESPRIT Prehistory

In 1978, on the basis of the "Europe + 30" report, the Commission created a working group, Forecasting and Assessment in the Field of Science and Technology (FAST). Under FAST, research institutes in the EC competed for grants to forecast developments in three main areas: work and employment, the information society, and the biosociety. The resulting thirty-six reports were synthesized by the twelve-person FAST staff at Science, Research and Development.[60] The FAST synthesis concluded that autonomous socioeconomic development in the EC would require "a coordinated ap-

60. Ros Herman, *The European Scientific Community*, 153.

proach in order to derive maximum value from the scientific, technological and industrial potential of the Community countries."[61]

Regarding IT the group also noted the impact of technological change in making collaboration appear attractive to firms and governments: Because of rapid innovation and market change, it was difficult to foresee which technological options would prove essential in the long run. The report states:

> No country in Europe has the capacity to cover such a spectrum of technological options, which is the only way to achieve a satisfactory degree of technological flexibility. . . . Hence there is a two-fold task for the Community: first, to achieve a certain collaboration and division of labour in order pursue jointly a wide range of technological options; and second, to carry out those specific programmes where there are evident advantages of scale.[62]

The report singled out semiconductors as an area in which cooperative action was necessary to meet the strategies of the United States and Japan. The threat from Japan and the United States would become one of the Commission's primary refrains in pushing for ESPRIT. Indeed, one EC official was quoted in late 1981 as saying that the Japanese 5G initiative was an acute challenge for European firms, "many of whom are managing to stay profitable only by technical linkups with the Japanese. But this is dangerous for the future of European industry."[63] In planning for an EC IT program, Commission officials also had in mind the Japanese model of precompetitive R&D carried out by industrial consortia.[64]

Commission preparation for ESPRIT began among those officials working on FAST. According to the FAST report, Roland Hüber of the FAST team "initiated the conception and coordination" of a research project on EC needs and activities in "long lead-time R&D in information technology" (LLTRD-IT).[65] This preparatory study, begun in 1980, lasted about eighteen months and led to the formation of the Joint European Planning Exercise in IT. The aim of this project was to "create the conditions for European collaboration in long lead time R&D . . . through the European Strategic Programme for R&D in Information Technologies (ESPRIT) and

61. FAST, *Eurofutures: the Challenges of Innovation*, xi.
62. Ibid., 74.
63. Tom Manuel, "West Wary of Japan's Computer Plan," 104.
64. Interview 22.
65. FAST, *Eurofutures*, xii.

[contribute] to the definition of the necessary complementary measures to promote a competitive European IT industry in the 1990s."[66]

The LLTRD-IT project commissioned studies that became ammunition for the Commission in persuading the IT companies that cooperation was necessary. For instance, the Battelle Memorial Institute and other consulting firms worked on a three-year LLTRD-IT study to identify long-term R&D needs in data-processing.[67] The array of studies marshaled by the Commission made it knowledgeable enough about the technologies and the industries to be credible.

Davignon came to the industry post at the Commission in 1979 already possessed of a desire not to see repeated in IT the same "tragedy" that had overtaken the European steel industry, a sector with which he was familiar.[68] Thus Commissioner Davignon became the patron of the small group working on IT at the Commission. Davignon protected them from "the bureaucrats who wanted to squash anything new or original."[69] He converted the small group within FAST into the Information Technologies Task Force (ITTF), an independent body under the Commission not subject to any of the existing directorates. Davignon argued that industry had to make a major contribution to any EC program.[70] In September 1979 he proposed that industry, governments, and the EC together work out a telematics strategy.[71] His ability to rally industry would prove to be the key to getting ESPRIT started.

In the meantime the Commission was slowly educating the Council on the need for EC action in IT. By the fall of 1979 the Commission (that is, the ITTF) had drafted a document on telematics strategy to submit to the European Council at its meeting in Dublin at the end of November.[72] The document was based on an "exhaustive

66. Ibid., 74.
67. Tom Manuel, "West Wary . . . ," 104.
68. According to a senior scientific counselor to the Commission during the entire period who also participated in the founding of ESPRIT. Interview 15.
69. Interview 22.
70. *Agence Europe,* 24 March 1979, p. 10.
71. *Agence Europe,* 29 September 1979, p. 12.
72. The European Council consists of the heads of state or government of the member countries. Their meetings are also referred to as EC summits. The European Council should not be confused with the Council of Ministers (usually referred to simply as the Council), which in principle consists of the foreign ministers. The specialized councils have the same legal status as and are technically equivalent to the Council of Ministers; these group ministers with specific portfolios, such as research, agriculture, finance.

general survey" by the Commission of national ministries of industry, posts, and research, which indicated a desire for the Commission to outline an approach. The paper submitted to the Council was titled "La société européenne face aux technologies de l'information: pour une réponse communautaire." It outlined six points for an EC telematics strategy:

1. Dispel resistance to innovation within the EC.

2. Create a European market for telecommunications and data-processing based on common standards.

3. Develop basic microelectronics technologies.

4. Create data-bank services that would be competitive on world markets.

5. Create a communications network linking EC institutions and national governments.

6. Develop a common position for world space and telecommunications organizations.

The Commission noted that telematics was a rapidly growing sector (over 15 percent per annum) and was "strategically important," having a "direct effect on the competitive position of numerous other branches of activity."[73]

The Council responded by asking for specific proposals in the areas of microelectronics and telecommunications. The Commission proposed a microelectronics program the following July.[74] The Commission had begun working in late 1978 with industry and research institutes to specify technical and performance goals.[75] Davignon asserted in announcing the document that major national programs in France, Italy, Germany, and the United Kingdom would not be enough to narrow the gap between Europe and Japan or the United States.[76] The Commission envisioned a budget of 100 million UAs over four years.[77]

Time dragged on. Though the Council neglected the Commission proposals, they won approval in other EC bodies. Both the Eco-

73. *Agence Europe,* 7 November 1979, p. 8; 26 November 1979, p. 9; 28 November 1979, p. 5.
74. *Agence Europe,* 18 July 1980, p. 6.
75. *Agence Europe,* 29 November 1979, p. 16.
76. *Agence Europe,* 18 July 1980, p. 6.
77. *Agence Europe,* 11 September 1980, p. 5.

nomic and Social Committee and the European Parliament endorsed the objectives and expressed concern that the financial means being discussed were too limited.[78] The research ministers finally approved a program in November 1981, though at only 40 MECU. The Microelectronics Program would fund 30 to 50 percent of project costs. The areas chosen for funding all related to IC design and production equipment: step-and-repeat equipment, electron beam for direct writing on wafers, plasma etching and deposition, test equipment, and CAD for VLSI circuits. The program was to promote cooperation among makers and users of the technologies, and between industry and the universities.[79] Over two years had elapsed since the Council responded to Commission prodding by requesting proposals for a microelectronics program.

Davignon and the Knights of the Roundtable

Within months of submitting his first proposals for an EC telematics strategy to the Council in late 1979, Davignon invited senior officials from the ten largest computer and telecommunications manufacturers in Europe to meet with him. The agenda for the February 1980 meeting consisted of discussing the possibility of a European telematics strategy.[80] Those executives in attendance[81] agreed in general with Davignon's assessment and accorded his notion of a European strategy a highly positive reception. Some of the industrialists felt that the Commission had underestimated the likely impact of vigorous Japanese development efforts on European and world markets by 1990. They expressed strong agreement that there was a need to support the development of basic microelectronics technologies and to develop common standards for data-processing and telecommunications. The companies agreed to submit specific observations on the Commission's proposals and to meet again in March.[82]

78. *Agence Europe,* 23 February 1981, p. 15; 1 May 1981, p. 13; 7 May 1981, p. 10; 9 May 1981, p. 7.
79. *Agence Europe,* 10 October 1981, p. 14; 9 November 1981, p. 5; 14 November 1981, p. 13; the Council regulation is in the *Official Journal,* no. L/376, 30 December 1981.
80. *Agence Europe,* 8 February 1980, p. 10.
81. Including representatives of Siemens, Nixdorf, CGE, Thomson, CII-HB, ICL, Plessey, Olivetti, and Philips.
82. *Agence Europe,* 11 February 1980, p. 10.

Davignon stepped up his campaign at the time the proposed Microelectronics Program was submitted to the Council (July 1980). In a briefing for industry executives, he argued that the choice for Europe was whether to be assistants to the United States and Japan in their strategies or to take an active role. He pointed out that Japan had caught up to the United States in IC technology and production in only four years, and did it with a lower government expenditure than existed in Europe as a whole. Davignon held out the goal of increasing Europe's share of semiconductor production from 6 percent to 12 percent by 1985. He also advocated support for the semiconductor-equipment industry, the opening of markets for telecommunications equipment, and common standards.[83]

In late 1981 Davignon invited directors of the twelve largest European IT companies (the Twelve) to "roundtable" discussions on the future of IT in Europe.[84] Davignon told the assembled executives that Europe faced formidable challenges from the United States and Japan, especially given the major new programs emerging in both countries. He argued that national programs were not the right answer, as Europe would end up reinventing the wheel many times. One participant (an independent senior advisor to the Commission on IT) paraphrased Davignon's message as follows: "If you remain alone, you will be subordinate to the Americans and the Japanese." Davignon told the Twelve that if they perceived a problem and if they wanted to cooperate among themselves, then he would find the funds for a major program. The company directors agreed with Davignon's analysis, but the general response was that they could not immediately commit themselves to a major cooperative program.[85]

After discussions within firms and some initial contacts between them, the Twelve met in a second Roundtable. Davignon reaffirmed his promise and told the companies to propose a program. A steering committee of the Twelve was composed to do so. The steering committee heard reports from outside consultants and, after reach-

83. *Agence Europe,* 11 September 1980, p. 5.
84. The companies were Bull, Thomson, and CGE from France; Siemens, Nixdorf, and Allgemeine Elektrizitaets-Gesellschaft (AEG) from Germany; General Electric Company (GEC), ICL, and Plessey from the United Kingdom; Olivetti and Societa Finanziaria Telefonica (STET) from Italy; and Philips from the Netherlands. These companies became known, and I will refer to them hereafter, as the Twelve or the Roundtable companies.
85. Interviews 15, 46.

ing agreement on the broad outlines of a strategy, constituted a number of expert panels to work out the specifics. The general strategy, defined by the steering committee working with Commission officials, included five areas as being appropriate for a European program: advanced microelectronics, advanced computing, software technologies, office automation, and integrated computer systems for industry. The expert panels were in place by the spring of 1982. Already the Commission was aiming at an initial phase of pilot projects to begin in January 1983, with EC funding of about 10–12 MECU.[86]

At this stage, in fact, the Commission made its first formal submission to the Council, on 25 May 1982. The proposed program was titled ESPRIT: European Strategic Programme for Research and Development in Information Technology. The Commission document had the purpose of starting the process of discussion.[87] It argued for a set of pilot projects, proposals for which would be submitted in the fall of 1982, with actual work to begin in January 1983. The proposed program met with generally favorable reactions from the national delegations. At the expert level the Research Council delegations agreed with the Commission analysis of the crisis facing European industry and on the need for swift action.[88]

The Research Council gave a favorable reception to the ESPRIT proposal at the end of June 1982. The president of that meeting, in summarizing the conclusions of the debate, noted that IT was "crucial for the future of European industry." The Research Council also concluded that an EC program was urgent given "the state of competition in world markets" and Europe's having fallen behind.[89]

Throughout this period the Twelve continued to meet, at the executive level in the steering committee (meeting about once a month) and at the technical level in the five working panels.[90] Through these efforts the overall work plan and the pilot phase took shape. One participant, a senior executive from a Roundtable company, de-

86. Interview 15; *Agence Europe*, 26 May 1982, p. 13.
87. CEC, *Towards a European Strategic Programme for Research and Development in Information Technologies*.
88. *Agence Europe*, 28 June 1982, p. 7.
89. CEC, *Communication from the Commission to the Council: On Laying the Foundations for a European Strategic Programme of Research and Development in Information Technology: The Pilot Phase*.
90. Interview 53.

scribed how at least one of the meetings worked. He was in charge of a one-day conference held at his corporate offices. Each company set up shop in a room, and the executives circulated among the rooms, working in pairs to specify projects on which they could cooperate. At the end of the day they had agreements on potential projects, which later became the pilot phase.[91]

For the technical panels the Twelve sent engineers and technologists. Initially, about 100 people were involved.[92] As companies other than the Twelve, as well as universities and research laboratories, became involved, this number rose. André Danzin, advisor to the Commission on IT, later said that creation of the work plan involved 400 technologists from industry, plus 150 from government and private laboratories and universities.[93] Their discussions quickly became open and lively, leading some industry executives to worry about what they might be "giving away."[94] Danzin said that being authorized to speak openly, the scientists "told all," like an "intellectual striptease."[95]

The Pilot Phase

All this activity led by August 1982 to a proposal for the ESPRIT pilot phase. The Commission document declared that the overall objective of ESPRIT was "to provide the basic technologies which European industry needs to be competitive with that of Japan and the USA." The paper then outlined the fundamental features of the program. The most important was that ESPRIT had to be "aimed at pre-competitive technology."[96] The term *precompetitive* is a vague one. Research can take place anywhere on a continuum from basic scientific research to product development and manufacture. The precompetitive condition was meant to keep ESPRIT away from work on actual products nearing the market. The program had to be precompetitive for several reasons. Some of the Commission's

91. Interview 1.
92. CEC, *Communication from the Commission to the Council: On Laying the Foundations*, 7.
93. Speech at a seminar sponsored by the Centre d'Etudes Prospectives et d'Informations Internationales (CEPII), 26 June 1987, at the Commissariat Général du Plan, Paris.
94. Interview 5.
95. Danzin, CEPII speech.
96. CEC, *Communication from the Commission to the Council: On Laying the Foundations*, 6.

reasons are somewhat abstract, dealing with larger questions of optimal research funding; other reasons are more pragmatic.

The Commission argued that in Europe long-term industrial research had been neglected for the sake of catching up with developments elsewhere. Thus, Europe had been unable to focus enough resources on long-term research to reach the critical mass of effort. "The result is a lack of punch in any major attempt to develop new technologies, which no one company or country can tackle alone." ESPRIT would remedy that defect. It would "reverse the current emphasis and focus it on longer-term technological research, establishing co-operation between industries across the European frontiers as well as co-operation between universities, research institutions and industry."[97]

A more pragmatic reason for the emphasis on precompetitive R&D was that the Treaty of Rome did not provide for the Commission to run industrial policies. Anything that looked like the promotion of specific firms and industries in the marketplace would raise thorny political and constitutional questions. Another good reason for limiting the R&D to long-term areas was that companies would be reluctant to share technologies that they were preparing to introduce on the market. Precompetitive R&D did not directly threaten any commercial interests.

Still, the definition of the boundary between precompetitive and competitive R&D was intentionally left vague. One Commission official told me the best definition of precompetitive that he had heard: "Anything two companies are willing to do together."[98] The Commission proposal made it clear that "nothing in the programme will prevent companies continuing to compete in the market with products and systems derived from the work."[99] Indeed, one early member of the ITTF informed me that they had in mind the Japanese model of interfirm cooperation for precompetitive R&D, then competition on the market.

Still, the Commission has made efforts to give the term *precompetitive* a firm juridical basis, even though these developments followed the pilot-phase period I am discussing in this section. The legislation finally establishing ESPRIT defines precompetitive R&D

97. Ibid., 6, 16, 17.
98. Interview 26.
99. CEC, *Communication from the Commission to the Council: On Laying the Foundations*, 17.

as that for which commercial possibilities remain five to ten years in the future, placing ESPRIT somewhere between fundamental research and applied industrial R&D.[100]

Additionally, in late 1984 the Commission adopted Regulation 418/85, which exempted from EC competition (antitrust) rules certain kinds of R&D collaboration. The regulation is worded so as to apply clearly to R&D agreements under the Commission's programs like ESPRIT. For instance, the exemption applies if the R&D is performed within a defined program. The exemption covers joint R&D "on products or processes and joint exploitation of the results of that R&D." The rule seeks to preserve market competition by limiting the duration of the exemption in cases where a joint production and marketing agreement involves partners whose combined share of the EC market exceeds 20 percent. Even in the new regulation, however, concepts and wording remain fuzzy.[101] In practice, the competition rules have been relaxed for firms developing products on the basis of Commission-sponsored research.[102]

The proposal for the pilot phase set out several key themes that undergirded Commission efforts throughout the evolution of ESPRIT and even through RACE. First, the Commission highlighted the need to fortify Europe's ability to compete with the United States and Japan. In the section on microelectronics the Commission noted: "Both Japan and the United States have a significant lead in VLSI technology. Submicron work is required to stay in competition with them." One of the applications areas chosen by the Roundtable planners was office automation. Regarding that sector the Commission wrote: "Office automation is expected to become the largest single Information Technology market. In the USA both IBM and Xerox each spend more on this subject than European industry and academia combined, and the Japanese fifth generation computer concept is also aimed at these applications."[103] This acute awareness of Japanese and American competition was shared by the national governments, as I showed in the first section of this chapter.

---

100. *Official Journal*, no. L/365, 31 December 1985, 4.

101. Alexis Jacquemin and Bernard Spinoit, *Economic and Legal Aspects of Cooperative Research: A European View*, 37–44.

102. See, for example, William Dawkins, "Brussels Moves on 'Know-How' Accords," *Financial Times*, 3 September 1987, p. 4.

103. CEC, *Communication from the Commission to the Council: On Laying the Foundations*, 10, 13; Annex II, 4.

A second key theme laid out in the pilot-phase proposal was, not surprisingly, that national programs would not suffice. "Given that there is a shortage of qualified manpower and other resources, which means that companies or governments individually cannot address all topics on a sufficient scale, the concentration of ESPRIT on pre-competitive research will enable the necessary critical mass to be reached in key areas." At stake were crucial new technologies, "which no one company or country can tackle alone."[104] As shown in the first section of this chapter, France at least was rapidly becoming convinced that national programs were insufficient, and the other major countries were also dissatisfied with the national-champion programs of the past.

Finally, a third important theme of the proposal was the need to overcome market fragmentation through common European standards. In fact, the Commission identified standardization as one of the two most important reasons for EC action, referring to "the relationship between R&D, standards and markets." In other words, the Commission linked R&D to standardization and market unification. "The importance of standards for the creation of a more homogeneous European home market in the IT sector is capital." The Commission also made specific mention of standards in the sections discussing the microelectronics, software, and office-automation components of the program.[105] Standards would remain a consistent part of Commission telematics initiatives.

I have highlighted these three themes (international competition, the inadequacy of national programs alone, and the need for European standards) at some length because they dominated Commission efforts to sell the governments on collaboration.

In its proposal to the Council the Commission was careful to clarify that approving the one-year pilot phase "will involve no commitment to any further phases."[106] Such a limited commitment was seen as unlikely to alarm member states.[107] In practical managerial terms, a pilot phase, to begin in January 1983, would give the Commission and the participants experience in the "complex and difficult issues" of running such a program. The Twelve had outlined a total of fifteen pilot projects, which were described in

104. Ibid., 6, 16.
105. Ibid., 10, 11, 13, 15.
106. Ibid., 19.
107. *Agence Europe*, 10 September 1982, p. 11.

some detail in the proposal. The Commission added one more, for the development of a system that would allow ESPRIT participants to communicate quickly and efficiently. All sixteen pilot projects shared three main characteristics: They could be started immediately; they were useful in their own right; and, though lasting several years, they could be evaluated at the end of the pilot phase on the basis of "milestones."[108] EEC funding for the initial projects would be 11.5 MECU, to be matched by the companies involved.[109] The pilot phase was also meant to encourage the participation of small companies.[110]

The pilot projects fell under the five main headings chosen by the Twelve. The first three were basic technologies, the other two were major applications areas: (1) microelectronics, (2) advanced information processing (AIP), (3) software, (4) office automation, and (5) CIM.

In Annex I to the proposal, the Commission laid out in some detail each of the pilot projects, including technological objectives, first-year review milestones, and subsequent review milestones. For instance, the first microelectronics project, "Advanced Interconnect for VLSI," specified the goal of developing a four-layer, metal interconnection technology at a level of one-micron features, capable of production utilizing CAD. The project would last four years, with two review milestones to have been achieved at the end of the first year. The first software project, called the "Portable Common Tool Environment," aimed at developing the hardware, software, and standard interfaces for software design. This project would allow European software designers to work with a common set of methods and tools. The project had nine first-year milestones and was the largest of the pilot-phase projects at 3.1 MECU.

Annex III of the proposal outlined the technical objectives of the full ESPRIT program. Each of the five technical areas was subdivided into a number of subprograms, the microelectronics area having the most specific subprogram objectives. The bulk of the microelectronics program was devoted to silicon VLSI technologies, but it also addressed compound semiconductor devices (like gallium arsenide chips), sensors, optical devices, flat-panel displays, and chip

108. CEC, *Communication from the Commission to the Council: On Laying the Foundations,* 21.
    109. Ibid., 25.
    110. Ibid., 19.

packaging. The AIP and software areas had less precise objectives, while the office automation and CIM applications areas indicated rather general themes in their subprograms.

The Commission felt confident enough of a favorable reaction to ESPRIT by member governments that it published an advance notice for participation in the pilot phase in the *Official Journal* of 18 October 1982.[111] The European Parliament voted in favor of ESPRIT and urged haste.[112] In early November the Council passed a "common orientation,"[113] with formal approval coming on 21 December 1982 at the requested EC funding level of 11.5 MECU.[114]

The Commission published an official invitation for tenders in February 1983, drawing over 200 proposals from 600 companies and institutes. The total value of all the proposals reached 50 MECU.[115] The Commission, with the help of a senior executive committee that included national IT officials, chose thirty-eight projects to receive funding. Not surprisingly, about 70 percent of the pilot-phase funds ended up going to the Twelve Roundtable companies that had prepared the program and were thus already in a position to submit relevant proposals.[116]

Preparing the Main Phase

The success of the pilot phase eased the way for the main program. Both companies and governments responded positively to the trial run.[117] The thirty-eight contracts had barely been signed when the Commission introduced a proposal for the ESPRIT main phase. The Commission submitted this proposal on 2 June 1983.[118]

The document reiterated two of the Commission's main arguments. First, major programs in the United States and Japan threatened Europe's future in IT.[119] The stakes were depicted as incal-

111. *Agence Europe,* 21 October 1982, p. 11.
112. *Agence Europe,* 30 October 1982, p. 10.
113. *Agence Europe,* 8 November 1982, p. 12.
114. CEC, *Proposal for a Council Decision Adopting the First ESPRIT,* 6. The pilot phase was passed in the form of a Decision, which was published in the *Official Journal,* no. L/369, 29 December 1982, 37.
115. *Agence Europe,* 5 April 1983, p. 7.
116. Sherry Buchanan, "Small Firms Find a Niche in EC's ESPRIT Program," *International Herald Tribune,* 21 March 1984, p. 9.
117. Paul Cheeseright, "How ESPRIT Will Help to Bridge the Gap," *Financial Times,* 28 March 1984, Survey, p. 8; David Fishlock, "A Shared Approach to Scientific Study," *Financial Times,* 12 December 1984, p. 12; Interviews.
118. CEC, *Proposal for a Council Decision Adopting the First ESPRIT.*
119. Ibid., 1–2.

culably high: "IT will be the driving force of economic growth for at least the rest of the century, and is therefore a *major factor in economic affairs*."[120] IT was the key to new products, processes, and services, and thus to increased exports and employment.[121] Second, the Commission naturally repeated its argument that fragmented national programs were too small and overlapped too much.[122] In addition, the Commission highlighted the support of industry:

> Community industry has acknowledged that, in order to reverse the trend of increasing reliance on importing technology, only *joint strategic long-term research planning* and the concentration of resources through the definition and funding of technology goals of common interest on a Community scale, can have a *good chance of redressing the situation.*[123]

Thus the Commission proposed the launching of the first five-year phase of an envisioned ten-year ESPRIT program.

The total budget would be 1,500 MECU, with half that amount coming from the EC. The sum proposed could appear almost negligible, argued the Commission, compared with the total EC industry investment in R&D in the sector, $5 billion per year. Some American companies by themselves poured $2 billion annually into R&D for IT. The Commission asserted that the ESPRIT program would bring the amount of resources devoted to long-term, precompetitive R&D into the same range as that found in the United States and Japan. Europe's main competitors were allocating between 5 and 10 percent of their R&D investment to precompetitive research; the 1,500 MECU from ESPRIT would comprise 6 percent of the total EC R&D effort in IT, raising the share of precompetitive R&D in Europe so that it was on a par with that in Japan and the United States.[124]

In addition to the budget the Commission proposed a staff of 150 and a 7.7 percent share of the total budget for staff and administration.[125] The proposal promised that the work program for ESPRIT's second five-year phase would be ready before the end of the fourth year of Phase I. The Commission also provided for an overall

120. Ibid., Appendix, 4, emphasis in the original.
121. Ibid., 7.
122. Ibid., Appendix, 20.
123. Ibid., 9, emphasis in the original.
124. Ibid., 4; Appendix, 44–45.
125. Ibid., Appendix, 62 and Financial Statement.

evaluation of the program at its mid-point, two and a half years into the work. The document projected a January 1984 launching.

Selling ESPRIT

The key to Commission success in winning approval for ESPRIT was that it did not try to do the job itself. It got industry to sell the program to the national governments. Quite simply, the Commission recruited industry, industry became committed to the cause, and industry took the cause to the national capitals. The launching of ESPRIT stems directly from the alliance struck by the Commission with a powerful transnational coalition, namely, the Twelve Roundtable companies. Of course, as I argued in the first section of this chapter, the governments were already in an adaptive mode, the failure of national-champion policies having led to a search for new solutions to the telematics crisis. This, in fact, was a necessary precondition. That the adaptive process resulted in collaboration is explicable only by the actions of an entrepreneurial IO allied with a potent industrial coalition.

Why were the Roundtable companies responsive to Davignon's invitation to put together a cooperative R&D program? I outlined in the previous chapter the major motives behind the rapid increase in interfirm alliances involving European companies in the early 1980s. These were the increasing costs of R&D, the blurring of boundaries between sectors (convergence), and the uncertainty induced by rapid technological and market change. The studies I cited in that chapter were based on extensive databases that recorded the motives expressed by businesses entering into partnerships. In other words, the motives I attribute to European firms that established technological alliances were not derived from some deductive schema but rather reflect the actual, empirical needs expressed by European telematics companies in the early 1980s. The Commission invitation presented a new way of meeting those needs.

In fact, there had been scattered talk about cooperation in European industry before Davignon's invitation.[126] The Commission acted as a seed around which European collaboration could crystallize. The Roundtable companies agreed to talk, and their dis-

126. According to an executive with a French Roundtable company, a British Roundtable executive, and a Commission official involved since the origin of the ITTF. Interviews 18, 28 and 60.

cussions entailed a learning process. According to all the participants I spoke to, it was through the meetings of the steering committee
and later the technical panels (as described previously) that the Twelve
discovered strategic areas in which they could fruitfully collaborate.
Those areas became the basis of the work programs. The loyalty of
the Twelve was ensured by having them prepare the pilot projects
and the work programs. In a real sense ESPRIT was their program.

Later, the companies were instrumental in selling the program
to the governments. The numerous studies the Commission had paid
for in working up to ESPRIT had to vanish. The Twelve had to
present ESPRIT to their governments as something built by their
efforts. If they had relied on Commission studies the Twelve would
have carried no credibility in their respective capitals. "That was
the key to selling the program to the governments."[127] The substantial work invested by the Twelve in the steering committee and
the technical panels thus gave ESPRIT an indispensable industrial
legitimacy. In its proposal to the Council the Commission quoted
extensively from a letter written jointly by the Twelve to Davignon.
The letter noted Europe's miserable market position and argued that
national programs even in the larger states would be insufficient.
The prescription of this transnational industrial coalition was equally
clear: "Unless a *cooperative industrial programme of a sufficient
magnitude can be mounted, most if not all of the current IT industry could disappear in a few years time.*"[128]

Industrialists and government officials in France, West Germany,
and the United Kingdom confirmed to me that the Roundtable
companies played an essential role in persuading their governments
that ESPRIT was a good idea. One French executive said that Davignon used the industrialists to convince the governments. The same
executive noted that the three French Roundtable firms (CGE,
Thomson, and Bull) presented a unified, clear view to the Ministry
of Industry, whose minister was persuaded and took up the cause
of ESPRIT.[129] Another French Roundtable industrialist, from a different company, said that he personally visited the Ministry of Research and Higher Education many times and lobbied the finance
minister as well. His opinion was that the Twelve had helped im-

127. Interviews 23 and 28.
128. CEC, *Proposal for a Council Decision Adopting the First ESPRIT,* 3, emphasis in original.
129. Interview 18.

mensely in winning approval for ESPRIT by lobbying their governments.[130] An official concerned with ESPRIT in the French Ministry of Research and Higher Education informed me that the industrialists did help convince the French government of the program's merits. As he put it, "When the three largest firms in electronics and information technologies are all saying it's a good idea, the officials listen."[131]

A similar phenomenon took place in Britain. Roughly the same people represented the British Roundtable companies (GEC, Plessey, and ICL) in ESPRIT as worked with the Alvey Committee. They therefore had a direct input into British policy-making circles. The United Kingdom's big three companies were definitely in favor of ESPRIT.[132] In addition, the senior civil servant placed in charge of the Alvey Directorate, Brian Oakley, also sat on the ESPRIT Management Committee from the very early stages. This arrangement permitted close coordination of the work of the EC program with that of Alvey. Oakley also became an advocate of ESPRIT within the British government. On the industry side, Derek Roberts, technical director of GEC, became persuaded at an early date that European collaboration was important.[133] As early as December 1982 Roberts was quoted as saying, apropos of ESPRIT, "There is simply no alternative to co-operation between previous rivals, and between industry itself and the academic world."[134] British industrialists became even more enthusiastic as the pilot phase progressed.

Finally, the German Roundtable companies also weighed in favorably with their government on behalf of ESPRIT. The BMFT held discussions with industry and industry associations regarding ESPRIT. German companies were in favor of the program. One BMFT official speculated that part of the industrial support for ESPRIT in Germany might have been due to reductions in BMFT industrial research funds at that time.[135]

In short, the picture painted by Commission officials, Roundtable executives, and national officials confirms the central role played by the Roundtable companies in winning national approval for ES-

130. Interview 23.
131. Interview 48.
132. Interview 47.
133. Interview 60.
134. Richard Brooks, "ESPRIT Puts d'accord into Europe," *Sunday Times,* 19 December 1982, p. 44.
135. Interview 4.

PRIT. Of course, final agreement did not come without political battles and compromises. In the rest of this section I describe the principal struggles. Note, however, that the disagreements centered on details of the structure and administration of ESPRIT, not on the fundamental merits of the program itself. The sales pitch of the Roundtable companies had reached national governments searching for new approaches to IT policy.

By summer 1983 it appeared that all ten member governments were in complete agreement on the importance and the content of the program. But signs of trouble were already appearing: Germany wanted to wait and see the results of the Stuttgart Mandate on the reform of EC finances. The Germans declared themselves willing to approve the new expenditures for ESPRIT only if they could be financed within the current budget allocations.[136] This would prove to be the main sticking point.

The Research Council meeting of 26 October 1983 brought forth a surprise however. French Energy Minister Jean Auroux, sitting in for Minister of Industry and Research Fabius, blocked any decision on ESPRIT. He rejected the funding level proposed by the Commission (750 MECU) and suggested a limit of 400 MECU. Auroux also asked for a provision limiting EC funding to genuinely European companies. His stand provoked puzzlement and confusion among the other delegations. For one reason, France had just the month before circulated its memorandum on increased technological cooperation in Europe. Furthermore, at the same time Auroux was applying the brakes to ESPRIT, French Minister of European Affairs André Chandernagor was advocating in the Groupe Unique de Préparation the development of an EC industrial policy to improve the competitiveness of European enterprises.

At the Research Council meeting seven delegations expressed agreement on the proposed level of finance. West Germany and the United Kingdom did not oppose the 750 MECU budget but withheld their approval pending the results of the Athens Summit scheduled for December, at which the overall EC budget problems would be addressed. Some of the smaller countries, especially Belgium, expressed the fear that they would not benefit from the results of the program because they had no giant firms like the Roundtable com-

136. *Agence Europe*, 24 October 1983, p. 7.

panies.[137] The ministers scheduled a special session for early November to resolve these differences.

At the Research Council of 5 November, Fabius at the outset lifted all the French objections. What explains the rapid turnaround? The problem stemmed from a French bureaucratic glitch. Apparently, when Auroux was detailed to substitute for Fabius, Auroux was briefed by functionaries who were not up-to-date on the government's thinking about ESPRIT and were unenthusiastic about the EC plan. Remember that the fall of 1983 was the period when the Mitterrand government was shifting to its pro-European stance. Auroux therefore thought he was representing the French position, but that position was in the process of being changed.[138] In the same meeting the Ten agreed on a budget of 700 MECU, though Germany and Britain continued to reserve final assent until after the Athens Summit.

Other organizational details were hammered out. First, the annual work program would have to be submitted to the Council, where it would be approved by qualified majority.[139] Second, Type A projects would receive 75 percent of the total budget, and Type B projects were guaranteed 25 percent of the total. Type A projects were defined as those with budgets of 10 MECU or more, later reduced to 5 MECU;[140] Type B projects would have no minimum. This was a concession to the small countries, led by Belgium, who wanted some assurances that small projects would not be overlooked.

Third, in special cases (namely, for universities and small- and medium-sized companies), the limit of 50 percent EC funding could be exceeded if agreed by qualified majority in the Management

137. *Europolitique*, 29 October 1983; 9 November 1983.
138. Interview 15.
139. Qualified majority is a voting procedure in the Council in which national votes are weighted and passage of a measure requires more than a simple majority of the weighted votes. Prior to 1986 the voting of the ten member states was weighted as follows: ten votes each for France, Germany, Italy, and the United Kingdom; five each for Belgium, Greece, and the Netherlands; three each for Denmark and Ireland; and a generous two for Luxembourg. Out of the total of sixty-three votes, forty-three were needed to pass a measure that required a qualified majority. When Spain and Portugal joined the EC, they received eight and five votes respectively; the qualified majority changed to fifty-four out of seventy-six votes. Qualified-majority voting in bodies other than the Council (like the ESPRIT Management Committee and the RACE Management Committee) follows the same format.
140. ESPRIT Review Board, *Mid-Term Review of ESPRIT*, A2.

Committee. Fourth, Type A projects had to be approved by qualified majority in the Management Committee. If the qualified majority could not be achieved but there was a clear majority opinion,
the Commission could make the final decision itself. The Germans
expressed a reservation on this point, fearing too great latitude was
being given the Commission. Fifth, the share for personnel and
administration in the ESPRIT budget was fixed at 4.5 percent, which
involved a concession from the Germans, who favored a 1 percent
limit.

Finally, the French relaxed their demand that ESPRIT funds go
only to genuinely European firms. They agreed to a statement in
one of the "considering" clauses that "those enterprises best placed
to achieve the strategic objectives of the ESPRIT program will be
recipients of assistance." Because the main ESPRIT objective was
to promote European industry, the clause indicated that European
firms would receive the grants.

The November meeting concluded with a political agreement on
ESPRIT, with final approval awaiting the outcome of the Athens
Summit. Commissioner Davignon expressed the view that if the
Athens meeting failed to achieve reforms that would free funds for
ESPRIT from other areas, the Commission would take money for
ESPRIT from its other R&D programs.[141] In the meantime, both
the Economic and Social Committee (unanimously, with one ab-
stention) and the European Parliament approved the ESPRIT pro-
gram.[142]

The Athens Summit failed to resolve the EC budget crisis. ESPRIT
thus remained blocked, even though all the members approved of
the technical content and the management provisions.[143] The Germans indicated an area of possible compromise: They would be
satisfied with guarantees that the money for ESPRIT could be found
in existing research appropriations.[144] The British did not express
any firm position. In fact, according to a *Financial Times* story the
British were in a holding pattern behind the Germans because of
divisions within Whitehall.[145] This speculation was confirmed to me

    141. *Europolitique,* 9 November 1983; *Agence Europe,* 7 November 1983,
p. 5.
    142. *Agence Europe,* 1 October 1983, p. 15; 15 October 1983, p. 11.
    143. *Europolitique,* 14 December 1983.
    144. *Agence Europe,* 14 December 1983, p. 10.
    145. Paul Cheeseright, "EEC Nears Information Technology Agreement," *Fi-
nancial Times,* 27 February 1984.

by a well-placed British civil servant. The cabinet was divided, with different ministries expressing contradictory positions in public.[146] Clearly, Thatcher herself had not yet decided in favor of ESPRIT.

At this point serious wooing of the Germans and British began. The French, who had the presidency of the Council for the first six months of the new year, initiated a series of high-level contacts with Bonn in order to reach an agreement. The Commission also pursued discussions with the holdout governments. Davignon himself made an appeal to the Council of Ministers meeting in late January 1984. He argued that the program could not await the next European Council, which was slated for March 19.[147] He warned the British and West Germans that their footdragging jeopardized Europe's efforts to catch up with the United States and Japan.[148] Eventually, Davignon traveled to London, where he met with Thatcher, and to Bonn.[149]

The Commission proceeded by publishing on 30 December 1983 an advance notice for participation in the ESPRIT main phase, so that interested parties would be informed about application procedures once the program was finally launched. Indeed, the work program had already been unanimously approved.[150]

At the Council meeting of 23 January the ministers agreed that a final decision on ESPRIT should be reached in the 28 February meeting of the Research Council. Contacts between the Commission and Bonn had produced a compromise solution. Germany would agree to the funding level of 700 MECU agreed on in December, provided that Commission R&D spending not exceed 600 MECU per year for the next two years. In addition, for that fixed amount, ESPRIT, the Joint European Torus, and the Joint Research Centers should have priority. This compromise would leave just enough money to keep other existing research programs going but limit some new projects. The smaller states objected to such a move because they derived benefit from the other programs but considered themselves less able to capture benefits from the three large programs.[151]

146. Interview 47; see *Agence Europe,* 10 February 1984, p. 7.
147. *Agence Europe,* 19 January 1984, p. 13.
148. Giles Merritt, "ESPRIT under Budget Cloud," *Sunday Times,* 22 January 1984, p. 25.
149. Jon Turney, "Cash Agreed for Europe's New Venture," *Times Higher Education Supplement,* 2 March 1984, p. 1.
150. *Agence Europe,* 19 January 1984, p. 13; 5 January 1984, p. 7; 25 January 1984, p. 7.
151. *Agence Europe,* 25 January 1984, p. 7.

Even so, the smaller states never argued against the ESPRIT program or its objectives. They did seek measures to assure that their organizations would be allowed full participation, as described previously regarding the Type A–Type B formula and the possibility of greater than 50 percent funding for some Type B projects.

By that point impatience was starting to mount. The European Parliament Energy and Research Committee passed an emergency resolution urging the Council to pass ESPRIT by 28 February at the latest.[152] There were vague rumors that the large IT companies themselves were getting fed up with the delays and had talked of setting up a program of their own, outside the EC. A program run by the major firms would naturally have reduced the role of small and medium enterprises (SMEs). In fact, the large firms were at least a little wary of including the SMEs, fearing that they would gain access to technology from the Twelve without contributing much in return.[153]

The Commission assented to the German conditions, promising to manage R&D expenditures so that ESPRIT would be run within existing appropriations in 1984 and 1985.[154] After that the Commission's own resources were scheduled to increase. This assurance satisfied the Germans and the British, and ESPRIT was finally released from its budgetary deadlock by the research ministers on 28 February 1984. Interestingly, the Council approved an ESPRIT budget of 750 MECU, the level originally requested by the Commission and higher than the compromise level of 700 MECU agreed to the previous December.[155]

Having prepared for this moment, the ITTF immediately published a call for proposals. By the deadline, it was inundated with 441 submissions. ESPRIT was taking off.

ESPRIT IN ACTION

This section examines ESPRIT from two different perspectives. First, how does it work, in organizational and procedural terms? Second, has ESPRIT been "successful"? The struggle surrounding the

152. *Agence Europe,* 4 February 1984, p. 12.
153. Sherry Buchanan, "Small Firms Find a Niche in EC's ESPRIT Program," *International Herald Tribune,* 21 March 1984, p. 9.
154. *Agence Europe,* 25 February 1984, p. 7.
155. *Europolitique,* 29 February 1984; *Agence Europe,* 1 March 1984, p. 7.

TABLE 7.2.   DISTRIBUTION OF ESPRIT I RESOURCES
AMONG SECTORS, IN MAN-YEARS

Subprogram	Year					Total
	1	2	3	4	5	
Microelectronics	186	258	360	410	456	1,670
Software	177	317	343	318	285	1,440
AIP	140	281	392	441	441	1,695
Office automation	210	310	440	390	100	1,450
CIM	121	216	215	220	172	944

SOURCE: Commission of the European Communities, as reported in "La commission a programmé le projet Esprit," *Les Echos,* 14 February 1984, p. 3.

launching of the second phase reveals much about the perceptions of ESPRIT on the part of participants, research ministries, and national governments.

The Mechanics of ESPRIT

The gears and levers to make ESPRIT work were already agreed on before the program received definitive funding. The work program projected that the resources (measured in man-years) would be allocated among the sectors as shown in Table 7.2. The R&D work would be precompetitive, and projects would as a rule have to include companies, universities, or laboratories from at least two different EC countries.

The Commission, through the ITTF, had overall responsibility for managing the program. Assisting the ITTF were ESPRIT's three other administrative bodies, the ESPRIT Management Committee (EMC), the ESPRIT Advisory Board (EAB), and the Steering Committee. The EMC represents the member governments, each state placing two members on it. As the states did not want to leave project selection entirely to the Commission, the EMC must give final approval to all projects and must specifically approve (by qualified majority) each project worth 5 MECU or more (Type A projects). The EMC must also approve any departures from official ES-

PRIT rules by simple majority—as in exceeding the 50 percent funding limit, for example. Voting in the EMC is weighted.[156]

The EAB is the participants' body. Its role is purely consultative, advising the Commission on technical questions. Its membership includes sixteen persons not on a representative basis but serving *à titre personnel.* About half the members come from Roundtable companies, the other half from small companies, universities, and laboratories. As one British Roundtable executive explained it, expanding the EAB beyond the Twelve was politically necessary; in order to win approval for the program, it could not appear to be merely a subvention for Europe's largest telematics houses.[157] The Twelve retain a central role through the ESPRIT Steering Committee (separate from both the EMC and the EAB), whose function is to outline the work plans every year. The work plans are then fleshed out by technical panels involving the Twelve, smaller companies, academics, and other researchers.

After the work program is approved and a call for proposals is published in the *Official Journal,* the ITTF begins the job of evaluating the proposals and making selections.[158] For this purpose the Commission retains technologists from industry, universities, and laboratories for a period of about three to six weeks. They serve personally, not as representatives of their organizations. Committees of experts evaluate the first part of the proposals, which at this stage are anonymous; the proposals lay out the technical objectives, show how they fit the ESPRIT program, describe the resources needed, and specify milestones (for checking progress) and deliverables. To fail at this stage, a proposal must be rejected by a majority of the experts on the panel. The committees then submit short lists with justifications of their choices.

The next stage is a study of the short lists by review committees each headed by a division head from the ITTF and including ITTF personnel and technical experts. At this point the proposals are no longer anonymous, as the committees must assess the experience and technical capabilities of each partner. The review also judges the statements of products and markets that might be exploited on

156. Arthur F. P. Wassenberg, *Strategies and Tactics of European Industrial Policy-Making,* 11.
157. Interview 60.
158. In the following paragraphs I rely on interviews with Commission officials and on ESPRIT documents, especially CEC, *ESPRIT: 1987 Information Package.*

the basis of the research. Each division head submits a short list. The final choice of projects is made by the head of the ITTF in conjunction with the division heads. Their choices then go before the EMC.

In practice the percentage of projects approved by the ITTF and rejected by the EMC is very low. The EMC does have the prerogative to suggest changes in projects. It is in the EMC that countries press for a good share of the work. Naturally, sometimes countries strike bargains to include one more partner on a project or to combine two projects. Nevertheless, according to one member of the EMC project selection has not been highly political and *juste retour* concerns have not posed a problem.[159] There is no formula to assure *juste retour,* and the Commission does not publish any figures on national shares. Significantly, organizations that have been involved in ESPRIT have not expressed any complaints about political meddling for *juste retour*. In a survey of participants conducted for the ESPRIT mid-term review, respondents did not complain of political interference in the selection process. Some did, however, complain that merging projects caused serious problems. But merging projects is not necessarily the outcome of political bargaining; it is also due to the desire to fund as many participants as possible on a limited budget.[160]

Once projects receive approval, the Commission signs a contract with each consortium partner. When the contract is signed, the consortium receives an advance of 30 percent of the Commission contribution. The partners are jointly and individually liable for fulfillment of the terms of the contract. One of the partners serves as the consortium leader (or manager) and acts as liaison for the project with the Commission. Within the ITTF a number of project managers keep tabs on an assigned set of projects. Each project manager must submit a monthly report on activities and progress. A larger evaluation involving outside experts retained by the Commission takes place every six months at the Commission, not at the work sites. Further disbursal of EC funds depends on satisfactory evaluations.

Regarding intellectual and industrial property rights, contractors in a project have equal ownership and equal rights to exploit the

159. Interview 47.
160. ESPRIT Review Board, *Mid-Term Review,* 17–19.

results, with specifics to be agreed on among themselves. Participants in different ESPRIT projects have privileged access to the results of another project if those results would enhance their ESPRIT work. Other EC companies can acquire the rights to knowledge generated in ESPRIT on a regular commercial basis. The Commission retains the right to require an ESPRIT participant to grant licenses if it does not wish to exploit results itself.[161]

Beginning with the pilot phase EC organizations responded enthusiastically to these arrangements. Out of 145 proposals submitted for the pilot phase, 36 were selected. As noted, the first call for proposals for Phase I attracted 441 submissions, including pilot-phase projects applying for continuation under the full program. A total of 110 projects received funding, including 23 carried over from the pilot phase. The total funding commitment for the 1984 call for proposals, after some changes in projects, settled at 372 MECU.[162] Of this first batch, exactly half were Type A projects and half Type B. The May 1985 call for proposals drew 389 responses, of which 95 were approved for a total of 279.7 MECU. Thus, by June 1986, after some settling out among the projects, 201 were underway with a total allocation of 677.7 MECU, or about 90 percent of the total 750 MECU budgeted for Phase I.[163]

A limited call for proposals followed in the fall of 1985, restricted to the software area and totaling 18.9 MECU. (The Commission had noted a less-than-expected level of participation in the software segment.) Out of over forty proposals, eleven received the stamp of approval.[164] Finally, the Commission published in April 1986 a final call for proposals for Phase I, with only 62 MECU left in the coffer. The office-automation area was excluded from this final round, the Commission judging that the program was already meeting its objectives in that sector.[165] Twenty projects were retained from this final round.[166] After the usual settling down and some attrition, the total number of projects in ESPRIT Phase I reached

161. CEC, *Proposal for a Council Decision Adopting the First ESPRIT*.
162. ESPRIT Review Board, *Mid-Term Review*, A2, n. 1.
163. CEC, *ESPRIT: The First Phase: Progress and Results*, 7.
164. *Europolitique*, 26 February 1986.
165. *Europolitique*, 19 April 1986. The totals from all the bidding rounds appear to total over 750 MECU because the termination or scaling down of some projects released a small amount of funds that could be added to the 1986 call.
166. *Europolitique*, 19 October 1986.

226. The number of projects in which each country was represented was as follows; there are no surprises here:[167]

United Kingdom	158
France	157
Germany	133
Italy	94
Belgium	52
Netherlands	46
Denmark	34
Ireland	26
Greece	22
Spain	22
Portugal	14
Luxembourg	1

A few numbers will sketch in the nature of the participation in ESPRIT I. By type of organization, participation was as follows: 62 small companies (fewer than 50 employees), 84 medium companies (50–500 employees), 181 large companies (over 500 employees), and 199 universities and research institutions.[168] The average number of partners per consortium at the end of 1986 was 5.1, though the average number of industrial partners per project was 3.5. This difference was due to the broad participation of nonindustrial organizations (universities and research institutes): Out of 201 projects as of December 1986, 150 included nonindustrial partners. SMEs (those with fewer than 500 employees) participated in 50.6 percent of Type A projects and 61.4 percent of Type B projects. They did over one-quarter of the work on 60 percent of the projects in which they participated.[169] Overall, SMEs had a role in 65 percent of the projects and received 14 percent of the funding. The Roundtable companies participated in 70 percent of the projects and received

167. CEC, Directorate General XIII, *ESPRIT: The Project Synopses,* vols. 1–8.
168. CEC, *ESPRIT: 1988 Annual Report,* 3.
169. CEC, *ESPRIT: The First Phase,* 78–85 and Tables 7–9.

50 percent of the funding, though their share has been falling over time.[170]

The average number of partners per project was inflated by a few extremely large projects whose primary aim was standardization. When standards are at issue, not being included can be costly. And, indeed, for effective standards, the more the industrial participants the better. The largest project (in number of partners) was "AMICE: A European Computer Integrated Manufacturing Architecture," which had twenty partners. A number of projects had nine or ten members, like the office-automation project "Standardisation of Integrated LAN Service and Service Access Protocols."

Assessing the Impact of ESPRIT

An evaluation of the technological and industrial results of the total ESPRIT program is impossible at this point. In any case, the technical impact of ESPRIT is peripheral to the arguments of this study, which focuses on the politics of international cooperation. Nevertheless, we can look at the three evaluations of the ESPRIT I program undertaken by the Commission: one at the mid-point as promised in its proposal, one near the end of Phase I, and one after its completion. I shall summarize the main conclusions of the reports as succinctly as possible, then analyze the politics surrounding ESPRIT Phase II. As noted, the launching of ESPRIT II reveals much about the perceptions of the program on the part of the Commission, industry, and governments.

Judging ESPRIT I   Some observers have claimed that the stated goals of ESPRIT have changed over time. For example, Arnold and Guy argue that technological advance has been replaced as the chief objective, with European collaboration becoming an end in itself.[171] This claim is only partially true. Certainly the Commission has shifted its emphasis to highlight ESPRIT's success in expanding IT collaboration in Europe and to tout that as a primary end. To be fair, however, the Commission could not possibly point to ESPRIT's success in narrowing technology gaps simply because it is far too soon to detect such changes. Both technological advance and the promotion of cooperation now appear as the prime goals of ES-

170. ESPRIT Review Board, *The Review of ESPRIT, 1984–1988*, 17.
171. See Arnold and Guy, *Parallel Convergence*, chap. 5.

PRIT, though sometimes one appears first and sometimes the other. The mid-term review, for instance, claimed that its charge was to examine how well the program was meeting its objectives; it listed the first objective as "the promotion of European industrial cooperation" and the second as "the provision of basic technologies." The third objective was to "pave the way for standards."[172] However, when the progress and results report reiterated the three main objectives of ESPRIT, it reversed the order of the first two and placed technology first.[173] The report of the Third ESPRIT Conference incorporated both orderings: In the introduction it placed "basic technologies" first; but in his speech the Commission official in charge of ESPRIT, Jean-Marie Cadiou, mentioned industrial cooperation first.[174]

The Review Board appointed by the Commission for its mid-term evaluation was composed of three outside experts: A. E. Pannenborg, chairman, formerly of Philips; André Danzin, formerly of Thomson and an advisor to the Commission on IT; and H. J. Warnecke of the Fraunhofer Institut/IPA. The secretariat to the Review Board also consisted of three outsiders: V. C. Grandis of Tecnetra, P. A. Monseu of Comase, and P. A. Walker of Mackintosh International. Their evaluation was based on personal interviews or group meetings with representatives of 131 organizations (participants and governments) and a questionnaire sent to 477 participants, of which 238 responded.[175]

The first conclusion of the review was that "there is now unanimous agreement as to the initial success of ESPRIT, particularly with respect to the promotion of trans-European cooperation." Respondents noted that cooperation added overhead costs (travel, coordination) of about 10 to 20 percent of project costs, but that nevertheless cooperative R&D was more beneficial than isolated efforts. There was "unanimous agreement" that ESPRIT had permitted the size, scope, and goals of research projects to be increased. The resources applied to projects, both manpower and diversity of skills, were greater than they could otherwise be. ESPRIT was also leading to increased confidence and cohesion within the European IT industry. On the whole, respondents thought that the

172. ESPRIT Review Board, *Mid-Term Review*, 5.
173. CEC, *ESPRIT: The First Phase*, 7.
174. CEC, *3rd ESPRIT Conference: Building Momentum*, 3, 37.
175. ESPRIT Review Board, *Mid-Term Review*, Executive Summary, i.

ITTF management of the program was exceptionally proficient. Several corporate officers mentioned that ESPRIT was at least as efficiently run as national programs with which they had experience. With regard to national programs, participants felt that they were complementary to ESPRIT, with no severe problems of competition.[176]

The main complaints regarding the program were that the application process was costly; there were sometimes delays in receiving contracts and payments; the merging of projects caused serious problems; the monthly reports were an onerous task; and the ESPRIT Electronic Information Exchange System was not at all adequate. National governments, then, would have some reason to accuse ESPRIT of being excessively bureaucratic, a rationale on the part of some states for supporting EUREKA (as will be seen).[177]

The questionnaire elicited a number of illuminating responses. I will highlight some of them in summary form:

1. Ninety-seven percent of respondents thought that project collaboration had been either "good" or "very good."
2. "Would you be undertaking this work without ESPRIT?"
   Yes, at a reduced level: 70.0%
   No: 24.1%
3. "Has ESPRIT led to more ambitious objectives?"
   A lot more: 16.9%
   Yes: 67.6%
4. "Has ESPRIT accelerated or slowed down your R&D (in terms of days)?"
   Accelerated: 29.9%
   No change: 36.8%
   Slowed: 32.9%
5. For 91.4 percent of respondents, European cooperation provided either "large" or "medium" benefits.
6. "Is there free information exchange between partners?"
   Excellent: 34.1%
   Good: 59.0%
   Little: 5.4%
7. "Will ESPRIT strengthen the European technology base?"
   Yes: 85.7%
   No influence: 13.0%
   No: 1.3%
8. When asked if they had to recruit new staff to carry out ESPRIT projects, 64.7 percent of respondents answered "yes."[178]

176. Ibid., 33.
177. Ibid., 43–49.
178. Ibid., Annex C.

The Review Board also made recommendations concerning the future of the program. Many of the suggestions were distilled from the comments of participants. These included:

1.   That ESPRIT II receive enough funding to increase substantially the amount of research and that the five areas be combined into three.

2.   That Phase II include large demonstration projects consisting of final systems and increased funding.

3.   That evaluation of proposals be streamlined and that greater attention be paid to "strategic and commercial significance."

4.   That project managers (prime contractors) be granted more responsibility.

5.   That communications be improved, including a newsletter and a better electronic communications system.

6.   That European "centers of excellence" be created.

7.   That continuity be assured by overlapping contracts and avoiding provisional contracts.[179]

Most of these suggestions were incorporated in the plans for ESPRIT II.

The progress and results report was based on 200 reports from the technical evaluators who conducted the six-month progress evaluations. It attempted to present to the Council "concrete technology results." The report noted that almost 90 percent of the projects were on schedule as of mid-1986, with fourteen projects three to six months behind schedule and only eight projects lagging more than six months. It summarized the positive role of ESPRIT as concentrating scarce resources, both personnel and finance; allowing the pursuit of more options; accelerating research; and taking advantage of other EC policies such as standardization and trade policy.[180] The bulk of the document consisted of numerous specific examples of technical milestones achieved in each of the five technical areas. I will summarize some of them for purposes of illustration.

In microelectronics a project produced a design for a 10K-gate array with two-micron line widths and 200 picosecond delay. Sie-

179. Ibid., 50–57.
180. CEC, ESPRIT: The First Phase, 2–8.

mens was building a 100 MECU production line for the arrays. Another project developed methods for multilayer metal interconnections in VLSI circuits, a technique exploited by Plessey on a 200,000-element complementary-metal-oxide-on-silicon (CMOS) chip. One major development in the software area was the prototype for a computerized software production system, the Portable Common Tool Environment, that could be run on computers of different makes. One company started marketing a commercial application of the system.[181]

In AIP, ESPRIT projects produced enhanced versions of the logic-programing languages (computer languages that run on logical rules rather than mathematical operations) used in AI work. A Belgian institute released a compiler for the Prolog language that produces faster executing code than any other in the world. Work in the office-automation area led to a technical standard for multimedia (voice, text, and image) office documents. The Herode project developed what became the European and world (International Standards Organization, or ISO) standards for office documents. The CIM projects also produced the bases for standards in manufacturing systems, such as factory communications and robot integration. Numerous large and small companies developed prototypes and marketable products from ESPRIT CIM projects.[182]

The 1988 annual report of ESPRIT reviewed further technical achievements obtained from projects. Without going into details, the report claimed that by the end of 1988, 130 of the 226 projects had produced 166 "major results." This output included forty-two direct contributions to products or services already commercially available, and forty-eight to products or services under development but not yet on the market. Thirty technical results had contributed to international standards.[183]

Like the mid-term review, the final review of ESPRIT I was carried out by a team of outside experts, again headed by Pannenborg. The Review Board based its conclusions on 210 interviews with participants, 949 questionnaires returned by participants, interviews with external consultants, and ESPRIT documents. Their report praised ESPRIT for substantially achieving its major goals but

181. Ibid., 25.
182. Ibid., 68–74.
183. CEC, *ESPRIT: 1988 Annual Report,* 3.

also frankly discussed the program's shortcomings. The Review Board concluded that "in the vast majority of projects trans-European co-operation has been a success and resulted in significant benefits for the participants." The report lauded ESPRIT for improving Europe's technology base, in both material and human resources, and for improving links between industry and the universities. On the whole, program management was "satisfactory and smooth."

Responses to the survey questionnaire showed that over 60 percent of participants thought that ESPRIT I objectives had been "well" or "adequately met" in AIP and CIM, but only about 40 percent responded similarly for the other areas (software, microelectronics, office systems). An additional 25–45 percent answered that ESPRIT objectives had been "partially" met, leaving a small portion (5–20 percent) who thought they had been "poorly" met. Ninety percent of respondents thought that collaboration with partners from other countries had worked "well" or "adequately." The most commonly reported benefit from ESPRIT work was "enhanced know-how" (70 percent of respondents), followed by "more ambitious R&D goals" (60 percent), "new products" (47 percent), "improved development techniques" (45 percent), and "improved products" (35 percent).

The major criticisms of the program were as follows:

1.   The early work plans did not address the technologies relevant to the core businesses of the major firms.

2.   The infrastructure for communications among participants was wholly inadequate, hindering access to results.

3.   In the key microelectronics area the goals were too ambitious for the funding available, and resources were spread too thinly over a range of technologies.

4.   The process of contract negotiation was too long in many cases, with insufficient feedback to proposers.

5.   The disbursal of funds was often delayed.

6.   The number of participants in a single project was sometimes too large, the practical maximum appearing to be six.

7.   The Commission should have more frequently halted projects that were not achieving results.

The Review Board noted that many of the criticisms had been taken into account while preparing ESPRIT II, and the second phase would therefore include a number of improvements.[184]

In my own interviews, British, French, and German executives from Roundtable companies all expressed general satisfaction with ESPRIT. The major benefit they cited was that the program gave them access to more results that would have normally been possible with their limited R&D personnel. Many company officials cited a new sense of self-confidence and community in the IT industries across Europe.[185] As I will argue later in this chapter and in succeeding chapters, the momentum built by ESPRIT Phase I propelled the EC forward into ESPRIT II and RACE and in other directions outside the EC framework. The new sense of "Euro-optimism" in IT thus preceded and foreshadowed the mounting enthusiasm tied to the 1992 movement. ESPRIT showed that Europe could act coherently to promote joint gains in crucial areas.

Perceptions of ESPRIT I and the Battle over Phase II   The growing reservoir of support for ESPRIT in industry and among research ministries did not mean automatic approval for the second phase. It was not sufficient because ESPRIT II became tied up in the debate over the Commission's proposed Framework Programme for R&D, of which ESPRIT was the largest element. The Framework Programme embraces all the R&D activities run by the Commission, including nuclear fusion, energy, environmental protection, and medicine. The governments of two states in particular—Germany and the United Kingdom—sought reductions in the proposed budget for, and increased control over, the content of the Framework Programme.

When it became clear that ESPRIT I funds would be fully allocated by early 1986, the Commission announced its intention to press for an accelerated launching of ESPRIT II. Officials in Brussels declared that in order to preserve the momentum achieved under the first phase, the second should begin in 1987, two years earlier than originally planned.[186] Meetings with the Roundtable compa-

---

184. ESPRIT Review Board, *Review of ESPRIT.*
185. Interviews 1, 5, 18, 37, 45, 46, 53, 54 and 60.
186. Peter Marsh, "EEC Bid to Boost Spending on ESPRIT," *Financial Times,* 18 November 1985, p. 3.

nies began in order to devise a plan for Phase II. The industry group recommended that the level of work be increased to 30,000 man-years, about triple the amount that the first phase would reach. This expansion would also require about triple the level of funding, or about 2.2 BECU from the EC.[187]

A first communication to the Council regarding ESPRIT II was completed in May 1986. The document stressed familiar themes. First, the economic importance of IT: "IT is not only a major industry in its own right but contributes significantly to the competitive status of most economic activities." Second, the Commission pointed out Europe's declining position in IT. It cited studies projecting a decline in the share of European producers in world IT markets from 23 percent in 1980 to 21 percent in 1990, by which time Europe would provide 30 percent of world demand, the largest single market.[188] Third, the document repeatedly highlighted Japanese and American programs, including SDI, that threatened to increase the technological lead of these countries over Europe.[189]

The Commission proposed a few major changes for ESPRIT II, though the basic attributes of the program would remain unchanged (precompetitive research, the system of proposals and contracts). I will describe these as quickly as possible. One big change was that ESPRIT II would be proposed within the context of the Framework Programme on R&D. Pursuant to the Single European Act (Luxembourg, December 1985) the Commission was to encourage industrial R&D in the EC. Unanimous approval of the Council was needed only for the Framework Programme as a whole; the specific R&D programs constituting it would be decided by majority vote. Thus, the Commission in 1986 was preparing the Framework Programme to cover the five years from 1987 to 1991. One highly placed Commission official within the ITTF told me that inclusion of ESPRIT II within the Framework Programme was a tactical decision. Technically, it could have been submitted individually. The Commission thought that attaching the big programs with broad support, like ESPRIT, to the Framework Programme

187. *Europolitique,* 7 December 1985. My conversations with industrialists confirmed that the enlarged goals for Phase II did indeed emerge from industry.

188. CEC, *Communication from the Commission to the Council: The Second Phase of ESPRIT,* 14–15.

189. Ibid., 4, 8, 10–11, 16.

would pull along the smaller programs that would not be approved if left on their own.[190] I will explain momentarily the difficulties to which this decision led.

Another significant change was that the Commission recommended that organizations from countries belonging to the European Free Trade Association (EFTA) be allowed to participate not just as subcontractors (as under Phase I) but as contracting partners. Non-EC partners would have to pay their own costs plus, when appropriate, a share of the operational expenses.[191] In addition the five technical areas would be collapsed into three: microelectronics, information-processing systems (including the old software and AIP sectors), and applications systems (including the old office-automation and CIM categories). More importantly, the Commission responded to Council desires for a more practical emphasis than that of ESPRIT I by proposing the inclusion among ESPRIT II projects of a small number of large-scale, focused technology-integration projects (TIPs). The TIPs would require the integration of ESPRIT systems and components results in applications relevant to EC users. The Commission also suggested creating European "centers of excellence" in basic IT research. Finally, the Commission proposed a set of measures to encourage technology transfer—that is, the diffusion of results throughout EC industry and users and especially to SMEs.

The total proposed by the Commission for the new, five-year (1987–91) Framework Programme (the overall plan for EC spending on R&D) was 10.3 BECU, which compared to the current Framework Programme of 3.7 BECU for the four years 1984–87. Of course, the new plan included major new R&D programs, notably ESPRIT and RACE. The Commission's figure was soundly rejected by the three large states, with the other nine members supporting the Commission. France, Germany, and the United Kingdom all argued that the program needed to set priorities, to be more selective, to take greater consideration of what was being done in national programs, and to include more rigorous evaluation procedures for present and future work. Britain's minister for IT, Geof-

190. Interview 67.
191. CEC, *Communication from the Commission to the Council: The Second Phase of ESPRIT*, 5, 41, 42. The EFTA countries are Austria, Finland, Norway, Sweden, Switzerland, and Iceland.

frey Pattie, said the 10.3 BECU budget was about twice as high as it ought to be.[192]

During the torturous struggle that extended over the following year, the value and importance of ESPRIT Phase II were never questioned by those states that resisted the Framework Programme. In attaching ESPRIT II to the Framework, Commission officials had run the risk of delaying their flagship program in the fight over smaller ones. A handful of programs, including ESPRIT, RACE, and BRITE (Basic Research in Industrial Technologies for Europe), were those that the French, Germans, and British saw as essential. They wanted other parts of the Framework Programme pared. Indeed, because specific programs needed only majority approval once the Framework passed unanimously, the large states used the initial battle over the Framework budget to try to control the list of programs that would later be included.[193]

Thus, it was an irony that ESPRIT was one program that the member governments vigorously approved of. In fact, the Council had passed a Resolution in April 1986 expressing a strong commitment to ESPRIT and urging the Commission to proceed with the program while implementing many of the changes suggested by the Review Board.[194] National officials in France, Germany, and Great Britain all told me either that their governments strongly supported ESPRIT II or that it (and RACE) were their top priorities in the Framework Programme.[195] The programs that were in doubt included the Community's Joint Research Centers, the Joint European Torus, and a number of smaller programs.

The battle focused on the budget, with Germany and Britain holding out for cuts. By spring 1987 bargaining had reduced the Framework Programme budget to 6.48 BECU, to which all states except Britain agreed. The British came under increasing pressure and even scorn from the Commission, the European Parliament, and other governments. In mid-April the European Parliament voted 141 to 4 to have the Commission withdraw the Framework Programme if the United Kingdom did not approve within three weeks.

192. Paul Cheeseright, "EEC Bid to Triple Research Spending Fails," *Financial Times*, 11 June 1986, p. 2.
193. Several of my sources made this point.
194. *Official Journal*, no. C/102, 29 April 1986, 1.
195. Interviews 4, 30, 47, 48 and 51.

Some delegates discussed the possibility of setting up a separate technology program among the other eleven countries.[196] Interestingly, Britain would be a net beneficiary of the research programs (receiving in work more than it contributed).

As national elections approached in the United Kingdom, it became clear that no decision would be taken until after the vote. The Commission was claiming that further delays for ESPRIT would cause irreparable harm to the IT sector.[197] Pressure was also mounting from industry. The heads of the Roundtable companies met with Commissioner Karl-Heinz Narjes in late June to express their concern. They argued that the delays already meant that 600 researchers would have to be released or reassigned temporarily until the fresh funds became available to continue ESPRIT projects. Such an outcome would compel the dismantling of some research teams, which the industrialists argued would alone constitute a serious technological and industrial setback.[198]

Indeed, industry was enthusiastically behind the second phase of ESPRIT. It was at the initiative of the Roundtable that the work program for Phase II was tripled over that of Phase I, to 30,000 man-years. Executives from IT companies in France, Germany, and Great Britain expressed to me strong approval for the second phase. No one spoke against it.[199] It would be difficult to argue that the prospect of subsidies inevitably disposed industry to look favorably on the program; the companies put up substantial earnest money for the program—namely, one-half the cost of each of their projects. German executives strongly favored an expanded ESPRIT II, one of them noting that it was much better to spend the money on R&D than on wasting tomatoes or piling up butter.

A French Roundtable executive declared that French industry presented a united front to the government in lobbying hard for ESPRIT II, and that his company would likely increase its ESPRIT involvement from 130 to about 200–220 researchers. A British executive told me that his firm was committed to supporting the increased funding for ESPRIT II with its own financial contributions.

196. William Dawkins, "UK Attacked on Research Funds," *Financial Times,* 10 April 1987, p. 2.
197. William Dawkins, "EC Research Ministers Scrap Talks on Cash Plan," *Financial Times,* 10 June 1987, p. 2.
198. *Europolitique,* 24 June 1987.
199. Interviews 1, 5, 18, 37, 40, 45, 46, 53 and 60.

Another, from a separate company, declared, "We have been doing our best to persuade the U.K. government to support the Framework Programme, especially including ESPRIT and RACE as essential."

The Electronic Engineering Association, Britain's largest trade association for the electronics industry, called publicly for British approval of the Framework Programme. The Association stated that further delays "must be detrimental to UK interests" and that 60 percent of its members participated in EC-sponsored R&D programs. The statement was timed for political leverage, coming out the week before an EC summit at which the Programme would be discussed.[200]

At the end of June, after the British elections that returned Thatcher, the new cabinet had two new ministers with responsibility for the Framework Programme, Kenneth Clarke and Lord Young. In contrast to Pattie, neither opposed it.[201] Finally, unanimous political agreement on a 5.2 BECU Framework Programme came from the budget ministers in late July 1987. An additional 417 MECU was held out at British insistence until there was evidence of progress in EC budget control. This compromise was formally ratified by the research ministers in late September 1987.[202]

As the Council was finally unblocking the Framework Programme, the Commission was submitting its formal proposal for ESPRIT Phase II. It had created a new organizational basis for doing so: In April 1987 the ITTF had been converted into Directorate General (DG) XIII for Telecommunications, Information Industries, and Innovation. DG XIII would administer the ESPRIT and RACE programs as the old ITTF had done. In the ESPRIT II proposal the original budget of 2.2 BECU for the Community contribution had been cut to 1.6 BECU, as even some of the national officials who were sold on ESPRIT thought the first figure too high.[203] The Commission once again cited the "substantial increase in public support

200. Terry Dodsworth, "Electronics Body Calls for End to EC Budget Delay," *Financial Times,* 26 June 1987, p. 7.
201. William Dawkins, "UK Close to Decision on EC Research Proposals," *Financial Times,* 24 June 1987, p. 2; William Dawkins, "Belgium Pushes for UK Backing on Research," *Financial Times,* 25 June 1987, p. 3.
202. William Dawkins, "Commission Approves Eight Research Projects," *Financial Times,* 23 July 1987, p. 2; William Dawkins, "EC Unblocks Long-Delayed Research Funds," 29 September 1987, p. 2.
203. Interview 47.

of funding of R&D in [the] USA and Japan."[204] The proposal outlined the research areas as follows:

1. Microelectronics and peripheral technologies
    a. High-density ICs
    b. High-speed ICs
    c. Multifunction ICs
    d. Peripheral technologies (displays, storage and retrieval systems, printers, sensors, etc.)
2. Information-processing systems
    a. Systems design
    b. Knowledge engineering
    c. Advanced systems architectures
    d. Signal processing
3. IT applications technologies
    a. CIM
    b. Integrated information systems (for office and home)
    c. IT applications support systems[205]

Phase II also included a basic-research component, with a 50 MECU budget.[206] Topics for research would include molecular electronics, AI and cognitive science, solid-state physics for IT, and advanced system design.

At the same time, the Commission presented its draft work program for the new phase. The plan had been prepared by representatives from 150 EC companies working in the technical panels.[207] Each of the three major areas constitutes a chapter, divided into the subheadings listed above. Each subheading is further divided into specific projects. Type A projects are outlined in some detail, including technological objectives, approaches, and time scales. Most subheadings contain one and sometimes two TIPs intended to combine technologies into usable systems slightly downstream from the precompetitive realm. The TIPs "aim at meeting ambitious, well-defined industrial targets."[208] The TIP projects specify milestones

204. CEC, *Proposal for a Council Regulation concerning the European Strategic Programme for Research and Development in Information Technologies (ESPRIT)*, 4.
205. Ibid., 16–23.
206. Ibid., 23–24; *Europolitique*, 24 June 1987.
207. CEC, *Proposal for a Council Regulation concerning ESPRIT*, 32.
208. CEC, *Draft ESPRIT Workprogramme*, 0–3.

and deliverables. For instance, the TIP for high-density ICs aimed at a complete integrated system for the design, production, and testing of application-specific ICs. A TIP in advanced systems architectures had as the ultimate goal a prototype parallel-processing high-performance computer. IT applications support systems included a TIP on a multimedia integrated workstation that would employ emerging technologies for workstations capable of handling voice, text, and image data.

Type B projects are not specified; rather, the plan lists within each subheading research topics considered appropriate for the small projects. The Type B themes consist of a general descriptive phrase—for example, under the CIM subheading, "strategies and tactics for advanced flexible automated assembly" and "advanced grippers and artificial skins," and fourteen others.[209]

After the adoption of the Framework Programme, each of its constituent programs had to be approved by majority vote in the Council. This approval came for ESPRIT II in April 1988. But the Commission had already, the previous December, published a call for proposals. The deadline for submissions was 11 April 1988, and by that date the Commission had received well over 1,000 applications. The following day the Research Council gave final approval to ESPRIT II at 1.6 BECU. Project selection had to be particularly severe: Only 156 of the proposals, involving 585 organizations, were granted contracts. The new basic-research program in ESPRIT also released its first call for proposals in spring 1988. The Commission received 300 proposals. Late in 1988, sixty-two proposals (285 organizations) were selected and eleven working groups were constituted, with a total budget of 63 MECU for thirty months. As noted, the basic-research program focuses on fundamental research on topics relevant to ESPRIT goals, including microelectronics, computing, and AI.[210] Table 7.3 shows national levels of representation in ESPRIT II and basic-research projects.

In short, the drawn-out struggle to win Council approval for ESPRIT II revealed solid, enthusiastic backing for the program among Europe's major IT firms. The Twelve proposed a tripling of re-

209. CEC, *Proposal for a Council Regulation concerning ESPRIT,* III-17.

210. CEC, *ESPRIT: 1988 Annual Report,* 3, 65–67; Terry Dodsworth, "Joint Bid for EC Computer Project," *Financial Times,* 11 April 1988, p. 34; William Dawkins, "Ministers Give Go-Ahead to High-Tech Research Projects," *Financial Times,* 12 April 1988, p. 2.

TABLE 7.3.   NATIONAL PARTICIPATION IN ESPRIT II
AND BASIC RESEARCH PROJECTS

	Number of Projects		
	ESPRIT II	*Basic Research*	*Total*
EC countries			
France	115	57	172
United Kingdom	119	53	172
Germany	107	49	156
Italy	74	34	108
Spain	74	17	91
Netherlands	58	26	84
Belgium	38	27	65
Greece	44	7	51
Ireland	32	14	46
Denmark	25	15	40
Portugal	27	9	36
Luxembourg	2	0	2
EFTA countries			
Sweden	11	9	20
Switzerland	8	4	12
Norway	7	4	11
Finland	6	3	9
Austria	3	2	5
Iceland	0	0	0

SOURCES: Commission of the European Communities, Directorate General XIII, *ES-PRIT: The Project Synopses* (Brussels, September 1989), vols. 1–8, *passim;* Commission of the European Communities, *ESPRIT: 1988 Annual Report* (Brussels, 1988), 65–66, 79–89.
    Note: Total number of ESPRIT II projects: 156. Total number of Basic Research projects: 73.

sources devoted to ESPRIT—in the end they were doubled—and were ready to match the EC contribution with funds of their own. The avalanche of proposals for Phase II also demonstrated an impressive degree of interest in the program on the part of IT organizations. The research ministries in the three large countries all favored the continuation and expansion of ESPRIT. Yet, in spite of such impressive backing, the program was held hostage for over a year to the opposition of a pair of states to a separate—though connected—issue, the Framework Programme.

SECONDARY EFFECTS OF ESPRIT

Will ESPRIT alter the structures of the IT industries in Europe? Will it make Europe a formidable competitor in world markets? No one can yet say. However, already the IT business in Europe is different. ESPRIT has produced ripple effects, or changes that have taken place outside the program but that are traceable to its influence. I will briefly describe two sets of phenomena growing from the program. The first includes concrete instances of technology cooperation that spun off from ESPRIT; the second is a less tangible but nevertheless widely recognized sense of community in European telematics industries.

Spinoffs

One indication that ESPRIT is having a durable impact on telematics in Europe is that a number of additional projects have begun as a direct result of contacts and cooperation established in ESPRIT. Indeed, RACE is a follow-on to ESPRIT. But ESPRIT also spun off other ventures that did not belong to any EEC or intergovernmental program.

In the spring of 1983 Bull, ICL, and Siemens began discussing the possibility of creating a joint laboratory for R&D on advanced computing. The idea was initially launched by Jacques Stern, president of Bull.[211] By September the three companies had agreed on a joint research center to focus on precompetitive R&D with full sharing of the results but no joint product development.[212] The European Computer Research Center (ECRC) officially came into being in January 1984. The ECRC has its own facilities outside of Munich and a staff of fifty full-time researchers. Its projects divide into four groups: computer languages, knowledge-based systems (expert systems), person/machine interaction, and symbolic computer ar-

211. Guy de Jonquieres and Paul Betts, "European Computer Makers Plan Joint Research," *Financial Times*, 22 March 1983, p. 44.

212. Guy de Jonquieres, "European Alliance in Computer Research," *Financial Times*, 2 September 1983, p. 34.

chitectures (as opposed to math-based architectures).[213] Executives at both Bull and Siemens traced the joint research center directly to ESPRIT and the increased contacts among the Twelve that it brought about.[214]

Another example of an ESPRIT spinoff is the merger of SGS Microelettronica (the semiconductor division of the state telecommunications company, Societa Finanziaria Telefonica, or STET) and the commercial semiconductor division of Thomson. The new SGS-Thomson, announced in April 1987, would be Europe's second largest producer of ICs, after Philips.[215] Thomson had been one of France's national champions in electronics, and its marriage to SGS signaled how far France had retreated from its ambitions for national autonomy in the *filière électronique*. An executive at Thomson told me before the merger was publicly finalized that he was absolutely certain that it came about as a result of ESPRIT interactions.[216]

Of greater significance than the industrial spinoffs of ESPRIT has been the acceleration of European standardization. Of course, a large share of ESPRIT projects addresses IT standards, and common specifications for future telecommunications systems have been a primary theme of RACE. But a handful of efforts to promote European standards in data-processing and computer networking have sprung up outside the EC programs. The first of these was the Standards Promotion and Application Group (SPAG). The group included the same twelve telematics giants that had begun meeting in the Commission's Roundtables in early 1982. The mobilizing initiative came from France's Bull. Stern, the new president of Bull, appointed by the Mitterrand government in 1982, was personally committed to European standardization for the computer industry. His goal was European adoption of the Open Systems Interconnection (OSI) standards for allowing computers of different makes to communicate and share programs and files. The OSI standards were (and are) being progressively defined by the ISO. Thus, Bull officials organized the first meeting of what became the SPAG group

213. Paul Tate, "Picking Up Speed," 64.
214. Interviews 18 and 45.
215. John Tagliabue, "Europe Joins Semiconductor Battle," *International Herald Tribune*, 12 May 1987, p. 19.
216. Interview 37.

in November 1982. One executive closely associated with the process at Bull said that SPAG was inspired in its origins by ESPRIT.[217] SPAG officially came into being in March 1983, with encouragement from the Commission. The group called on national governments to require conformance to OSI standards in public procurement.

At bottom SPAG was an effort to combat the domination of IBM. IBM had its own proprietary standard—Systems Network Architecture (SNA)—for linking computers. OSI would, its backers hoped, open the market by assuring customers that they could buy from any maker equipment compatible with any other machines. European firms hypothesized that many users bought IBM equipment so as to ensure that it would run on SNA with their previous IBM purchases. Thus, SPAG made specific proposals for European standards (based on the ISO models) in such areas as packet-switched data networks and message handling to European standardization bodies (Comité Européen de Normalisation, Comité Européen de Normalisation Électronique, and CEPT). SPAG submissions have been accepted as the basis for future standards work in Europe.[218]

SPAG efforts have generated other results. The British government announced in mid-1984 that it would start requiring that computers purchased by the government conform to OSI standards. In September 1984 IBM Europe announced that it would start developing products based on OSI.[219] After these initial victories part of the SPAG group created a joint company, SPAG Services S.A., in October 1986.[220] Based in Brussels, with an initial budget of 2.4 MECU per year, SPAG Services offers test facilities for verifying compliance with European standards. It therefore supports and demonstrates the interconnectability of European IT products.[221]

217. Robert T. Gallagher, "Stern Spells Europe's Future O-S-I, Not I-B-M," 43; Interview 18.

218. Eric Le Boucher, "L'Europe Informatique," Le Monde, 16 March 1984, p. 1; Herbert Donner, "The OSI World, Seen from SPAG Europe." Donner was chairman of SPAG Services S.A.

219. Richard L. Hudson, "IBM Europe Backs a Computer Language Pushed by Its Rivals," Wall Street Journal, 2 May 1986, p. 1.

220. The initial shareholders were Bull, ICL, Nixdorf, Olivetti, Philips, Siemens, STET, and Thomson.

221. Jean-Jacques Chiquelin, "L'Informatique Européenne Rentre dans la Norme," Liberation, 3 October 1986, p. 12; Donner, "The OSI World."

The second major standards initiative in IT addressed the operating system for computers. The operating system is the set of internal instructions in a computer that manages the flow of information within the machine and the execution of programs. This time the initiative came in 1984 from Robb Wilmot, then of Britain's ICL. The initial discussion involved five of Europe's computer makers (all of them Roundtable companies). An ICL executive informed me that the process of proposing the group to other partners was greatly eased by the technical contacts created in ESPRIT.[222] In February 1985 six firms (Bull, ICL, Nixdorf, Olivetti, Philips, and Siemens) announced the formation of the Open Group for Unix Systems. The objective of the group was to encourage the use of the Unix operating system developed by AT&T's Bell Labs, and it later became known as the X/Open Group.[223]

Again, the idea was to create an alternative to SNA. The thinking was that if a large group of computer makers committed to Unix, it would open the market for their products. First, software writers would be more willing to write programs that could be used on a variety of makes of computer. Second, buyers would be more willing to buy the computers knowing that a large body of software was available to run on them. In addition any programs written to conform to Unix would be operable on any machine built to the standard. Unix would provide operating systems for the fast-growing minicomputer and personal-computer markets.

In time, other makers joined X/Open, including Ericsson of Sweden and the European subsidiaries of DEC and Sperry of the United States. The X/Open Group has produced a manual that provides software producers with guidelines for writing programs according to the Unix applications environment. Initial victories came as some governments (France, the Netherlands, and Sweden) established Unix as a national standard, requiring it for public procurement orders. Significantly, IBM announced that it would offer a version of the Unix operating system on equipment ranging from personal computers to mainframes. A battle over which version of Unix would become the standard has since emerged in the United States, but X/Open agreed on the System V version. The Commission later

222. Interview 5.
223. "Six Constructeurs Européens Choisissent un Logiciel d'ATT," *Le Monde*, 19 February 1985; *Europolitique*, 20 February 1985; Robert T. Gallagher, "Europeans Are Counting on Unix to Fight IBM," 121.

granted an exemption to the X/Open Group from the EC ban on agreements among firms because any standards produced by the group would be published and would therefore not distort competition.[224]

These spinoffs provide tentative indications that ESPRIT is changing the way the telematics industries are organized and do business in Europe. At a minimum the collaborative programs have accelerated developments that would have occurred anyway but probably only after further delays and footdragging. ESPRIT provided an organizational structure in which the proper contacts could be made and a consensus could be fashioned. Prior to ESPRIT European firms sought out American companies for technology partnerships. Because of ESPRIT European companies now seek out European partners. In fact, Cadiou, the Commission's director of the ESPRIT program, pointed out that in 1983 (the year before ESPRIT was launched) there were thirty-two alliances linking European to American firms compared with only six between European firms. In 1986 forty-six intra-European linkups almost matched the forty-nine created with U.S. companies.[225] In addition, although European standards would likely have emerged in time, there is no doubt that ESPRIT hastened the process.

Less Tangible Effects

ESPRIT triggered a burst of collaborative activity in Western Europe and set off changes that are potentially more profound than those in organizations and procedures. The program has altered fundamental perceptions about European telematics. The best way to summarize those shifts in perceptions is to speak of the newfound sense of a European telematics community. Most of the industrial officials interviewed for this book spoke enthusiastically of a new appreciation of Europe's technological strengths and a fresh set of informal ties linking companies and technologists across the

224. Gallagher, "Europeans Are Counting," 121–22; *Agence Europe,* 20 December 1986, p. 11; Don Clark, "Computer Giants Gang Up on AT&T-Sun," *San Francisco Chronicle,* 17 May 1988, pp. C1, C4.

225. Louise Kehoe, "Seven Take a United Stand against AT&T," *Financial Times,* 18 May 1988, p. 27. The number of linkups with Japanese firms remained constant at eight.

continent. The new awareness and associations could alter the way European firms conduct technology and product planning.

Officials at ICL in the United Kingdom, for example, told me that the benefits of cooperation were far more important than the funding received: "We have new contacts and new confidence." One official noted that there was, after ESPRIT, more of a European IT community. Another executive declared that the main success of ESPRIT had been the creation of an IT network in Europe: "I can call my friends at Siemens or Philips and ask what's going on. This is hard to quantify but very important." A research director at GEC remarked that although his firm identified real technical benefits in software standards and semiconductors, there was "almost more benefit from the companies knowing each other better, to take advantage of the potential that was spread around Europe."[226]

A French Roundtable executive expressed a similar notion, noting that ESPRIT brings together some "2,000 or so European scientists and technologists, creating a European community of researchers." Another French industrialist declared that one of the spinoffs of ESPRIT had been extensive visiting of other laboratories, which allowed people to know better what was going on technologically elsewhere and so make better decisions about alliances.[227] More recently, Stern, chairman of Bull, argued that European cooperation in IT should go well beyond precompetitive R&D. He said that European firms are now more willing to collaborate than they were previously because of the success of ESPRIT and progress on common standards.[228]

Officials at Siemens recognized that ESPRIT allowed them to take advantage of the "synergies" of working with the other partners, from whom they had learned a great deal. An executive at Nixdorf noted a new sense of confidence in European IT firms that resulted from ESPRIT.[229]

In short, regardless of whether Europe closes any technology gaps because of ESPRIT and its follow-ons, both telematics policy and business practice have changed. Cooperative industrial ventures, new

226. Interviews 5, 53, and 60.
227. Interviews 37 and 68.
228. Alan Cane, "High-Tech Warning to Europe," *Financial Times*, 8 June 1989, p. 2.
229. Interviews 45 and 54.

standards initiatives, and the fresh sense of a telematics community imply that some of the changes will be durable.

CONCLUSION

ESPRIT confirms the utility of the theoretical framework outlined in Chapter 2. First, it addresses the question of demand for cooperation. Theory must account for the cognitive process by which state leaders come to choose cooperation, especially in an area as dominated by nationalist policies as IT had been. The cognitive-process variables were clearly at work in ESPRIT: The national-champion strategies of the past decade had failed to close any technology gaps. Moreover, European national policy-makers recognized the failure and were in an adaptive mode, searching for new means of promoting their IT industries. The evidence for this adaptative mode consists of major studies of the IT sector in France, Germany, and the United Kingdom, all leading to new and enlarged national support programs.

Second, analysis must account for the political leadership necessary to organize cooperation. ESPRIT shows that IOs can exercise that leadership. Indeed, the Commission was indispensable in ESPRIT. It both mobilized and shaped the European consensus behind collaboration. The Commission was technically well prepared, was headed by an entrepreneurial leader in Davignon, and faced an opportunity for leadership because of the policy crisis in member governments. The most important political factor in ESPRIT was the alliance struck by the Commission with the Roundtable companies. The Twelve designed the program and sold it to their governments. No political explanation of ESPRIT could suffice without assigning importance to EC institutions and to the transnational industrial alliance.

ESPRIT also incorporated organizational features that enhanced the prospects for successful collaboration. For one, the program was large. ESPRIT I was the largest EC R&D program to date, and ESPRIT II accounted for fully one quarter of the 1987–91 Framework Programme. The total budget for both phases was over $5 billion. The program was substantial enough in the first phase to include 420 organizations from across the EC. With that many pieces of work, chances were far better that each state could receive an acceptable share than had there been only a few dozen pieces to fight over. Furthermore, in the EMC states could bargain somewhat for the inclusion of national organizations in ESPRIT projects. Still, no participants pointed to EMC give-and-take as a problem for ESPRIT. Participation is voluntary, left up to the individual companies, universities, and institutes. Though each member country contributes to ESPRIT through the general budget, no country can complain of being forced to participate in projects it finds useless. Thus, ESPRIT satisfies *juste retour* concerns via a modified form of à la carte.

ESPRIT might not propel Europe to the same level as the United States and Japan in IT. But it has gone far in creating a European IT community. At a minimum ESPRIT was a bold political initiative, one that ended two decades of exclusively unilateral, nationalist IT policies in Europe.

EIGHT

RACE

Making Connections

With ESPRIT on track and out of the station, the Commission began to fire up the engine for its other telematics vehicle, a program in telecommunications. Its handling of ESPRIT had won for the ITTF a degree of credibility as an administrator of large-scale, industrially important, high-technology research. The time was right to capitalize on its favorable image and on the nascent optimism regarding European telematics collaboration that it had inspired. The telecommunications program, eventually called RACE, would follow the same pattern that had made ESPRIT a success: Start with an industrial coalition, then win the approval of governments. The recipe was a good one, though more complicated in the telecommunications sector.

As in semiconductors and computers, Europe's telecommunications-equipment industry was the domain of national champions. In fact, the same companies were sometimes prominent in both IT and telecoms, companies like Siemens, Philips, and GEC. A complication for the telecommunications sector was that it had long been the administrative demesne of the public telephone administrations in each country. The PTTs, with their historic monopolies on all aspects of telecommunications in Europe, added a new dimension to the Commission's task in RACE. In the end the PTTs approved of RACE after reshaping it somewhat to their liking. The key analytic factors in explaining the movement from crisis to collaboration are the same as those that figured in ESPRIT: an entrepreneurial IO, a powerful transnational industrial coalition, and national governments in the midst of profound policy adaptation. In the present chapter I first describe the crisis in telecommunications facing European governments. I then detail the political processes that led to RACE.

TURMOIL IN TELECOMS

The telecommunications realm has been in upheaval since the late 1970s. Much of the tumult stems from the microelectronics revolution. New electronics technologies have led to increasingly capable and efficient communications equipment, which makes possible myriad new services, which in turn strain the capacities of traditional institutional arrangements—namely, the PTTs. All these developments have been, and continue to be, exhaustively analyzed and commented on. I will therefore summarize as succinctly as possible the key technological and regulatory changes that have placed pressure on Europe's traditional telecoms structures.[1]

A Changing World

Telecommunications involves the creation of electronic links between two locations for the purpose of sending and receiving information. In early voice telephony, the most basic of telecommunications services, an operator sitting in front of a switchboard established the connections by plugging a jack into a socket. After World War II telephone connections were created by electromechanical switches, in which an electrical impulse triggered the movement of a metal reed. Progress in microelectronics in the 1970s

---

1. The revolution in telecommunications has been so dramatic that virtually innumerable books and studies on the subject have emerged. Most of what follows is general background knowledge, though I use primarily the following sources: Borrus et al., *Telecommunications Development;* Gilbert-François Caty and Herbert Ungerer, "Les télécommunications: nouvelle frontière de l'Europe"; Hart, "The Politics of Global Competition"; Jill Hills, *Deregulating Telecoms: Competition and Control in the United States, Japan and Britain;* Nguyen, "Telecommunications"; and OECD, *Telecommunications.*

made possible the development of fully electronic switches. In these switches the connections are made within the ICs themselves, involving no moving parts. Software programs control the switching. Thus, at one level today's telecommunications exchanges resemble computers: They are made of thousands of ICs and are controlled by preprogrammed series of instructions. Such "digital" switches can operate on digital or analog networks.[2]

New transmission modes have also revolutionized communications. Microwave relays permit transmission without the need for laying cables. Satellites can perform a range of telecommunications functions by making point-to-point connections, linking several locations, or even broadcasting (point-to-multipoint communications). The emerging generation of transmission facilities capitalizes on fiber optics. Glass filaments carry information in the form of pulses of light. The advantages of fiber optics are enormous: greater capacity (a single filament can carry three times more voice conversations than can a current coaxial cable), immunity to electromagnetic noise and interference, and lower rates of signal attenuation (meaning fewer repeaters). Plus, the fibers are made of one of the earth's most abundant substances, silica.

The new switching and transmission techniques have made possible a proliferation of services beyond basic voice conversation. As computers spread throughout business and industry, the need to transmit vast quantities of data ballooned. New data-processing companies, like Electronic Data Systems, depended on telecommunications links to receive inputs and to return data after processing them in their large mainframes. Numerous industries now require constant, reliable, high-volume data communications—for example, banks clearing their accounts or airlines processing their worldwide reservations and ticketing. Other new services include

2. Electronic switching is usually called digital because all the commands for running the exchange exist in the form of binary electronic bits—the same basis on which computers operate. A digital switch can, however, route traffic through an analog telephone network. In such "space-division switching," the transmission facilities do not carry digitized information but rather analog signals, which vary continuously over a range and mirror the modulations in the original message (for example, a voice). The digital switch simply completes the circuits through which the analog signals flow. Time-division switching involves a fully digital network: The original signals are converted into packets of digital information, which are fed into the network and routed through the switch unchanged. They are then reconverted if necessary into an analog form (voice, music) at the receiving end.

paging, mobile phones, videotex,[3] teleconferencing,[4] and electronic mail.

Up to now the different kinds of services have required physically distinct networks. For instance, telex, a text-transmission service, uses lines distinct from the telephone network. Presently, voice telephony and low-speed data transmission flow through the basic network. But high-volume data transmission and teleconferencing require circuits with greater capacity. Thus, users with extensive need for rapid data transmission frequently (in the United States, at least) use separate, fully digital networks. In the visual domain cable television employs a completely distinct network.

In the near future that will change. The next step in network evolution will probably be something that is now loosely called ISDN: Integrated Services Digital Network. ISDN will be fully digital (switching and transmission) and will integrate voice, data, and text services. Further in the future is broadband communications, which will likely be based on optical fiber, microwave, and satellite transmission. The increased bandwidth of broadband systems is like additional lanes in a freeway: It can carry more traffic at once. Broadband systems will be able to carry more bits of digitized information; whereas ISDN will carry two Mbit/s (two million bits per second), broadband telecoms should be capable of at least 140 Mbit/s, with some systems currently being designed for 600–1,440 Mbits/s. Broadband systems will permit a single network to carry everything from basic voice conversation to high-speed data to high-quality moving pictures (HDTV).

As technology made possible ever more advanced uses of telecommunications, the demand for new equipment and services took off, especially among major business and industrial users. As Bar and Borrus put it, "In ways never before possible, companies are able consciously to design and build telecommunications networks that decisively enhance their competitive position."[5] The burgeon-

---

3. Videotex is the name for a range of information services aimed at the mass (residential) market. The services are sometimes divided into teletext and videotext. Teletext is one-way transmission of text, such as news headlines or stock-market quotations. Videotext combines text and graphics and is interactive—that is, the user can interrogate the information service and give instructions. For example, with a personal computer and a modem a home user can now go on electronic shopping sprees, ordering from a videotext "store." Or the home user can peruse airline schedules and purchase a ticket herself. Videotext systems are frequently a gateway to dozens of different services. The best-known American videotext systems are probably The Source, CompuServe, and Prodigy.
4. Teleconferencing allows meetings among groups at geographically distant locations via linkups that transmit both sound and video pictures of the participants.
5. Bar and Borrus, *From Public Access*, 1.

ing demand among businesses for new telecoms services has placed strains on traditional institutional and regulatory arrangements. The pressures erupted in the 1980s into a continuous debate, in public as well as trade forums, over deregulation and liberalization. Policy adaptation was universal. In Chapter 6 I described how aggressive regulatory change in the United States and Japan exerted powerful pressures on Europe's traditional telecoms structures. Freed from past restrictions, IBM and AT&T entered new sectors (telecoms and computers, respectively) and made Europe a prime target for expansion. Japan's well-organized efforts to prepare for broadband communications threatened to place Japanese firms in the lead in the future for equipment and services.

In short both technological and regulatory changes were sweeping across the advanced countries by the early 1980s. Europe's PTTs could not find refuge in their entrenched monopolies. The national governments were (and are, and will be) adapting their telecommunications policies and regulatory structures. The Commission's proposals for European cooperation in managing the changes thus arrived at an extremely opportune moment. Everybody was preparing for an overhaul of telecommunications, but the precise lines of future networks (ISDN, broadband) and institutions could not yet be discerned. The Commission proposals therefore entailed not an upsetting of otherwise stable arrangements but a coordinated, regionally planned management of the changes underway. Of course, that did not mean there would be no disagreements over the direction and rate of adjustment. But national leaders in an adaptive mode were more receptive to new ideas than they otherwise would have been.

Adaptation in Europe

In telecommunications Europe in the early 1980s was not a community but a collection of fiefdoms. Each national telecoms fiefdom was ruled by the administration of posts, telegraph, and telephone. The acronym *PTT* reveals the origins and nature of the administrative structure: Telegraphs had been added onto the postal services, and telephones were an extension of telegraphs. Telecommunications was seen as a natural monopoly and a public utility. Because telecommunications was considered a natural monopoly, it was believed that only a single supplier of networks and services could achieve economies of scale. Thus, European PTTs were public monopolies of telecoms networks, terminals, and services. Telecommunications constituted a public utility because, as with roads and electricity, there were social gains from providing universal service. Opening segments of the telecommunications system to competition cuts against the public-utility ideology that has dominated telecoms policies in Europe.

Naturally, the national telecoms administrations resist any such encroachments on their domain, the kind of defense of bureaucratic turf one would expect of any organization. In addition the postal services in most European countries are heavily unionized, and telecommunications revenues have long underwritten postal losses. Thus, the postal unions resist anything that might cause layoffs or cut budgets.

Furthermore, it is by no means clear that deregulation American style is socially or economically optimal in the long run. Deregulation has led to a plethora of different public and private networks in the United States, providing advanced services to that part of the business community that can afford them. At some point the fragmentation may prove a handicap. With much of the telecoms realm in the United States privately owned by large users, innovation may be privatized. The existence of diverse networks may spread out revenues so much that no one can make the huge investments needed for network modernization (toward broadband, for instance).[6] By contrast, a single, universal network—guaranteed by a state administration—could carry the same services for business and make them compatible with each other but also provide access to advanced services for residential customers.

No one can predict unequivocally which will be the best route to liberalization and modernization—wide-open laissez faire or PTT-led evolution. Because of Europe's administrative heritage, the PTTs will have a major say in planning the transition to next-generation telecommunications. The reforms have begun, and in each country they follow a distinct path. Table 8.1 provides a chronology of the major events in European telecommunications policy.

The United Kingdom   Of the European countries the United Kingdom has proceeded farthest along the path of liberalization. The government commissioned in 1976 a review of the operations

6. These concerns are explored in ibid., and in Borrus et al., *Telecommunications Development,* 12–15.

TABLE 8.1.   TELECOMMUNICATIONS CHRONOLOGY

Year	Event
1972	France: CGE produces the world's first fully digital exchange.
1977	United Kingdom: The Carter Report urges the separation of the postal and telecoms functions of the BPO.
1978	France: *Plan télématique* begins.
1979	EEC: In September Davignon proposes an EC telematics strategy.
	EEC: In November the Commission submits its first document on telematics strategies to the Council at its Dublin meeting.
1980	EEC: In July the Commission submits its first specific telecoms proposals.
1981	United Kingdom: BT is created, and Mercury is licensed.
	France: CGE and Thomson are nationalized.
	France: Minitel is introduced on a trial basis.
	Germany: Work begins on the experimental wideband ISDN system BIGFON.
1982	United States: A modified judgment requires the breakup of AT&T.
1983	France: The telecommunications business of Thomson is merged with that of CGE in Alcatel.
	France: An experimental broadband system is launched.
	Germany: Videotex is introduced on a small scale.
	EEC: In September the Commission proposes six lines of action for EC telecoms strategy.
	EEC: In November industry and telecoms ministers agree to the creation of the SOGT.
1984	United States: IBM and AT&T are freed to compete in telecoms and computers, respectively.
	United Kingdom: The government sells 50.2 percent of BT stock to private investors.
	EEC: For ten weeks in the summer industry and the research arms of the PTTs meet in the PET, producing a proposal for RACE.
	EEC: In July the Commission signs a memorandum of understanding with CEPT.
	United Kingdom: In the fall the experimental ISDN network IDA begins operation.
	EEC: In November industry ministers approve measures for the coordinated introduction of new services and the first steps toward opening markets.

TABLE 8.1.   (CONTINUED)

Year	Event
	EEC: In December industry ministers approve the objectives for an EC telecoms strategy submitted by the Commission.
1985	Japan: In April a law is passed requiring reregulation of NTT.
	Netherlands: The Steenbergen Report urges liberalization of telecoms.
	Italy: A ten-year plan for reorganization and liberalization of telecoms goes into effect.
	Germany: The Witte Commission is appointed to study telecoms reform.
	EEC: In March the Commission proposes a RACE Definition Phase.
	EEC: In June the Commission submits proposals for standardization in IT and for mutual recognition of type approvals.
	EEC: In July the Council approves the RACE Definition Phase.
1986	EEC: In February the Commission announces the Definition Phase projects.
	EEC: In May the Commission submits proposal for the coordinated introduction of ISDN.
	EEC: In June the industry ministers approve the Directive on mutual recognition of type approvals.
	France: In September the CNCL is created to regulate telecoms and broadcasting.
	EEC: In October the Council approves the Special Telecommunications Action for Regional Development (STAR) program for telecoms development in the less-favored regions.
	EEC: In October the Commission proposes the RACE Main Phase.
	EEC: In December the Council approves a Decision on standardization in IT and telematics services, as well as a recommendation on the coordinated introduction of ISDN.
1987	France: In January the purchase of ITT by Alcatel is finalized.
	EEC: In February the Commission submits proposals on pan-European, digital mobile communications.
	EEC: In June the Commission publishes the Green Paper, and the Council approves a Directive on pan-European mobile communications.

TABLE 8.1.   (CONTINUED)

Year	Event
	France: An experimental ISDN system is established in Renan.
	Belgium: The Wise Men Report urges liberalization and reform of the RTT.
	EEC: In July the Council approves the Framework Programme.
	Italy: IRI-STET proposes reorganization of the fragmented telecoms system.
	Netherlands: Parliament adopts the Steenbergen recommendations.
	France: In September the VAN market is opened to competition.
	Germany: In September the Witte Report recommends moderate liberalization.
	EEC: In December the Council gives final approval to the RACE Main Phase.
1988	EEC: In January work begins on RACE Main Phase contracts.
	EEC: The Commission issues the Directive to end PTT monopolies on the supply of terminal equipment.
	EEC: ETSI is created.
	EEC: In June the Commission issues a second call for proposals for RACE.
1989	Netherlands: In January PTT Telecommunication is split from the postal service, and competition is opened in VANs and terminal equipment.

of the BPO, a public corporation that at the time encompassed both postal and telecommunications services. The Carter Report, which resulted (1977), recommended that the postal and telecoms activities be split into two different public corporations, and that the telecommunications network be modernized by rapid introduction of the System X electronic exchange. Thatcher's Conservative government acted quickly on the Carter recommendations, creating a nationalized company in 1981, British Telecom (BT). In December 1984 the government sold 50.2 percent of BT stock in a public offering. The new government also initiated in 1981 a study on the telecoms monopoly of BT. Following the Beesley Report, which resulted from the study, the government began to liberalize the markets for telecommunications networks, terminals, and services.

The first step was the authorization of a new carrier (network operator), Mercury, in 1981. The new company, now wholly owned by Cable and Wireless, laid fiber-optic cables along British Rail tracks, linking major cities in a fully digital network. Mercury aimed initially at the business market (especially in downtown London) and trunk (intercity) and international communications. Though Mercury remained small compared with BT (total revenues for 1987–88 of about £380 million versus £9,880 million for BT), the threat of competition forced BT to rationalize and formulate new strategies. Competition on network provision also remained limited because the government could not authorize any new carriers before the end of 1990.[7]

Liberalization touched other important areas of British communications. A new Office of Telecommunications (Oftel) regulates the industry (somewhat as the FCC does in the United States), ensuring competition and fair rates. The market for terminal equipment has been open since 1984, meaning that users can purchase telephones, modems, PABXs, and other devices from competing suppliers. Previously BT had a complete monopoly on customer-premises equipment. Even in switching equipment BT has introduced competition in its purchasing practices. BT has made clear that it will no longer buy solely from the traditional U.K. suppliers, Plessey and GEC (which merged their telecoms interests in April 1988) and STC. In fact BT has already placed some orders with Thorn-Ericsson and an AT&T-Philips joint venture, Pye TMC.

To modernize, BT has announced plans to merge its voice, data, and telex networks into an ISDN. The goal is to have 80 percent of customers connected to ISDN by 1992.[8] An experimental ISDN, called Integrated Digital Access (IDA), began operating in the fall of 1984. In preparation for ISDN, interexchange lines were fully digitized by 1989. One potential problem is that Britain has chosen an ISDN model different from that favored by other European countries and the CCITT.[9] The British expect to upgrade their sys-

7. Bar, "Telecommunications in the United Kingdom," Appendix, 76–79.
8. "Grande-Bretagne: des 'Points d'acces,' " 49.
9. The CCITT, a body of the International Telecommunications Union, formulates recommendations on international telecommunications standards. The British system comprises one channel for voice and data at sixty-four Kbit/s plus two channels for simultaneous data and signaling at eight Kbit/s each. The CCITT proposed

tem to fit the CCITT standards, while the current standard allows early use.

British deregulation has placed pressures on continental telecoms authorities. For instance, international telephone calls are significantly cheaper from London than from most other European capitals.[10] As a consequence, some firms are relocating their communications centers to London. The DGT in France had to adjust its tariff schedule, raising local rates to reduce international charges, in order to keep customers from routing their transatlantic calls through the United Kingdom.[11]

France   French telecommunications has followed the typical French high-technology policy: state-engineered corporate mergers and ambitious plans. The DGT was until 1987 the French agency charged with responsibility for telecommunications policy; it falls under the Ministry of Posts, Telegraph, and Telephone but generates its own revenues and has a budget independent of the national budget. The DGT, through its research arm, the Centre National d'Etudes des Télécommunications, pioneered research on fully electronic exchanges and in 1972 began installing CGE's E-10 switch, the world's first fully digital exchange.

The DGT also had a hand in industrial policy for the telecoms sector until 1987. The DGT folded ITT and Ericsson subsidiaries into a telecommunications division for Thomson in 1975. The Socialist government nationalized Thomson and CGE (the other main telecoms equipment supplier) in 1981, and two years later merged the telecommunications divisions of both companies into Alcatel, a CGE subsidiary. The DGT opposed that move because it ruined the policy of maintaining competition among suppliers. The DGT also protested when the government diverted a major share of its revenues to fund the *filière électronique* and to aid the national budget. All of a sudden the DGT lost its surpluses and had to borrow in order to make necessary investments. Alcatel created the world's second largest equipment maker by purchasing ITT's European telecoms subsidiaries in a deal finalized in January 1987. Later, the

---

standard includes two voice and data channels of sixty-four Kbit/s each and a single sixteen Kbit/s signaling channel.

10. Sabine Delanglade and Eric Rohde, "PTT: Déréglementation accélérée en Europe," *La Tribune,* 26 November 1985, p. 6.

11. John Wilke, "Can Europe Untangle Its Telecommunications Mess?" 47.

French government sold a small equipment firm, CGCT, to a group
including Ericsson (Sweden) and Matra (France). Finally, under the
government of Jacques Chirac, responsibility for the telecoms industry was transferred out of the DGT to the Ministry of Industry.[12]

From 1975 to 1980 the DGT presided over the aggressive expansion of the telephone system to bring the penetration rate up to
the level of other advanced countries. It sponsored in 1978 the *Plan
télématique,* which aimed at the rapid introduction of advanced telematics (combining computers and communications) services. The
Plan included experiments with a public videotex system, an electronic directory, broadband optical-fiber facilities, and plans for a
new telecoms satellite. Transpac, a packet-switched network for data
transmission, came on line in 1979.

The French have been among the most ambitious in Europe in
planning for advanced services and future networks. In 1987 their
level of digitization of the network (55 percent of switches, 70 percent of transmission) was the highest in Europe.[13] France has had
a pilot ISDN project since 1987 at Renan, and by 1988 the telecoms
administration was preparing the transition from its separate specialized networks (the four "Trans" systems) to a single ISDN network.[14] An experimental broadband system has been in operation
since 1983 in Biarritz with some 1,500 subscribers.

One of France's great successes has been the Teletel/Minitel system, a videotex service aimed at the mass market. The system offers
the DGT's Electronic Directory Service (an on-line telephone directory for the whole country) as well as the Kiosque. Private companies offer information services through Kiosque; by April 1986
1,900 data services were available to professional or residential customers. Use of the Minitel by the public has skyrocketed, largely
because of the policy of the DGT to "lend" Minitel terminals free
of charge. The terminals were introduced on a trial basis in 1981,
and by the end of 1985 1.3 million were in use.[15] Through December 1987, 3.37 million terminals had been installed.[16]

12. Thierry Vedel, "La 'Déréglementation' des Télécommunications en France,"
no page numbers.
13. "France: Le Pays le Plus Numérisé," 49.
14. See Marie-Laure Théodule, "La gamme Trans en pleine mutation vers le
RNIS."
15. Jeffrey A. Hart, "The Teletel/Minitel System in France," 21–25.
16. Paul Betts, "A Frenzy of Alliances," *Financial Times,* 11 May 1988, European Telecommunications Survey, p. 10.

Further changes liberalized several other aspects of French telecommunications. In September 1986 the Chirac government created the Commission Nationale pour les Communications et les
Libertés (CNCL). The CNCL was supposed to regulate telecommunications and broadcasting much as the FCC does in the United
States; formerly the DGT both ran and regulated the system. Also,
the CNCL could authorize private networks for internal use. In September 1987 the government published a new law opening up the
provision of VANs to competition. The law did not permit competition in basic voice telephony, which would remain the province
of the DGT. Consortia began to form to offer VANs, including one
joining IBM, Crédit Agricole, and Paribas. The next step was to
open competition in the provision of mobile telephone systems; two
groups were authorized. The Chirac government also renamed the
DGT France Telecom and spoke of plans to privatize it. But the
Mitterrand electoral victory in the spring of 1988 and the opposition of the Socialists and the unions brought an end to these plans.
The second Socialist government replaced the CNCL with a Conseil
Supérieur de l'Audiovisuel (CSA). The CSA has the responsibility
of authorizing large private networks and new VANs. A decree of
May 1989 created the Direction de la Réglementation, which func-
tions much like the American FCC; its charter was fully worked
out in 1990.[17]

France, during the period of reorientation after the failure of the
early Mitterrand project (see Chapter 7), became a vigorous promoter of European collaboration. The government announced its
intention to "work for the creation of a European telecommunications space." In a lengthy article published in Le Monde the minister in charge of the PTT, Louis Mexandeau, called for increased
European collaboration and expressed support for the Commission's telecoms initiatives. His statement stressed reciprocal market
opening and common European standards.[18] The shift in France
from national-autonomy strategies to a European outlook helped
ease the way for the Commission's proposals leading to RACE.

Germany   The Bundespost in the Federal Republic of Germany was for a long time Europe's most stubborn defender of tra-

17. Benjamin Coriat, "Régime réglementaire, structure de marché et competitivité d'entreprise."
18. Louis Mexandeau, "Pour une politique européenne des télécommunications,"
Le Monde, 3 April 1984, p. 35.

222   RACE

ditional PTT roles and monopolies. To be sure, the basic communications law of 1928 and the Fundamental Law of 1949 granted
the Bundespost the exclusive responsibility for building and running
telecommunications networks. The Bundespost has interpreted this
responsibility to include the networks, services, and even terminal
equipment.

As late as the mid-1980s private networks were permitted only
for in-house use and could not be connected to the public network.
The Bundespost retained its monopoly on the sale of the first telephone handset to every customer and until 1988 was the sole supplier of modems. The Bundespost even forbade the sale of computers with internal modems. Any company that wished to sell
terminal equipment directly to users (rather than to the Bundespost)
had to have each product approved by the Bundespost. For its network equipment purchases the Bundespost relied on a small circle
of German firms headed by Siemens, then SEL (the ITT subsidiary
now owned by Alcatel), along with Nixdorf and IBM Germany for
data communications and VAN equipment.

The Bundespost developed specialized networks to try to meet
the growing demand for enhanced services. The Integrated Digital
Network includes data transmission, packet-switched data, telex,
and teletex. In 1985 the Bundespost was pushing for broad expansion of Bildschirmtext, a videotex system, apparently with only
moderate success. In particularly close collaboration with Siemens
the Bundespost developed plans for ISDN. Because of a major error
in exchange development strategy, Germany lagged well behind in
digitization of the public network, the first digital switches being
installed only in 1984. The public ISDN, integrating voice and data
services, had nevertheless been installed in thirty-nine cities by 1990
and is to be completed nationwide by 1993. The next step will be
to integrate narrowband and wideband ISDN (adding videophone
and videoconferencing to the public network). Eventually, a universal broadband system based on fiber optics will be in place, adding radio and television to the public network. Trials were underway in 1985 for wideband ISDN in seven major cities under the
Bigfon project; the next step was to link the local networks via
fiber-optic trunk lines in the Bigfern project.[19]

19. Patrick Cogez, "Telecommunications in West Germany," 54–56; "ISDN
Makes Strides in West Germany," 22.

Pressures for liberalization from German business and pressures for increased openness to trade from the United States began in the mid-1980s.[20] The Kohl government appointed the high-level Witte Commission to study telecommunications reform, and it began its work in early 1985. The report finally emerged in September 1987. The Witte Commission majority recommended a set of reforms that fell short of full deregulation or privatization. In fact four members of the Commission issued a separate opinion, arguing that the proposals did not go far enough and that "only replacement of the monopoly with competition at all levels can lead to a market capable of withstanding the future."[21] The most important recommendations were these:

1.   To split telecommunications activities from the postal services and place them with a national enterprise to be called Telekom

2.   To have Telekom retain a monopoly on basic telephony and to open all other services to competition; to allow Telekom to compete with private firms in offering VANs

3.   To permit Telekom to lease lines for the offering of VANs to third parties; if it did not permit competitive leased lines, to authorize alternative networks after three years

4.   To align tariffs more nearly with costs, ending subsidization of local calls by long-distance calls

5.   To let customers choose the terminals they wished to attach to the network; to allow competition with Telekom on terminal maintenance; and to allow Telekom and others to set prices of terminals without submitting them for approval[22]

Moderate as they were, the proposals drew intense criticism. Only the Free Democrats strongly supported liberalization. Some Christian Democrat *Länder* governments objected to the reforms, even though Chancellor Kohl committed himself to seeing them through. The Social Democrats and Greens opposed the changes. In this position they supported the postal workers' union, which represents

20. Hart, "The Politics of Global Competition," 187; Peter Bruce, "Bickering Bonn Tackles Bundespost Monopoly with Reluctance," *Financial Times,* 2 June 1987, p. 2.
21. "From Bundespost to Telekom," 407.
22. "RFA: Les recommendations (en substance) de la commission Witte," 91.

463,000 of the Bundespost's 500,000 employees (making the Bundespost the largest single employer in Germany). The union feared "tens of thousands" of layoffs, rising tariffs, and diminished quality of services. Even the Christian Democrat coalition partner from Bavaria, the Christian Social Union, was reluctant.[23] Still, the changes were approved by the cabinet in May 1988, after court challenges by the unions failed. The Witte reforms became law in July 1988.

Elsewhere in Europe   Telecommunications reform was not limited to the three major countries of Europe. Italy's fragmented system felt the first breaths of rationalization. A government agency, Azienda di Stato per i Servizi Telefonici (ASST, under the Ministry of Posts and Telecommunications), runs the trunk lines. SIP, part of the state-owned STET conglomerate, manages most of the local networks and linkups to customers. A third company, Italcable (also part of STET), handles international connections. The Italian National Plan for Telecommunications (for 1985–94) foresaw open competition for VAN provision and for the supply of customer premises equipment. In 1987 the leadership of the Istituto per la Ricostruzione Industriale (IRI) group (which owns STET) proposed reorganizing STET into an operating company resembling BT. The new STET would include SIP, Italtel, Italcable, the satellite division, and an alliance with a foreign equipment manufacturer.[24] AT&T formed an alliance with Italtel, but the rest of the plan has not been realized.

The Netherlands, with the most tightly regulated telecommunications system in the EC, also moved toward competition. The Steenbergen Report of 1985 contained a number of deregulation proposals that found their way into a bill passed by parliament in 1987. Under the new law, which took effect on 1 January 1989, the postal services of the PTT and the telecommunications services were split into separate subsidiaries of a new state-owned enterprise, NV PTT Nederland. PTT Telecommunications retains its monopoly on network provision but must compete with private com-

23. Peter Bruce, "Bickering Bonn Tackles Bundespost Monopoly with Reluctance," *Financial Times*, 2 June 1987, p. 2; and David Goodhart, "Reforms Firmly on Track," *Financial Times*, 11 May 1988, European Telecommunications Survey, p. 10.

24. James Buxton, "Good Progress after a Swift Change of Direction in Italy," *Financial Times*, 24 October 1983, Survey, p. 12; Alan Friedman, "A Crucial Year," *Financial Times*, 11 May 1988, Survey, p. 12.

panies in an open market for terminal equipment and VANs. The Dutch are also planning to convert the network to ISDN after 1995 and to broadband by the turn of the century.[25]

The report of a Wise Men Commission in Belgium in 1987 recommended changes similar to those planned in the Netherlands. The Wise Men recommended that the RTT (Régie des Téléphones et Télégraphes) be converted into an independent state-owned enterprise. Under legislation proposed in 1988, the RTT will retain a monopoly over the network and over "essential" VANs (though these remained to be defined). The market for terminal equipment will be opened to competition. Belgium also began an exploratory technical study for broadband networks.[26]

From these snapshots of situations in various European countries, it is clear that telecommunications in Europe was undergoing profound alterations. The changes were both technological, as networks moved toward ISDN and broadband, and institutional, as the PTTs found that they would have to share some of their traditional markets with private competitors.

THE COMMISSION AND TELECOMMUNICATIONS

Into this unsettled atmosphere the Commission launched its trial balloons for European cooperation in telecommunications development. If national telecoms policy-makers had not been in an adaptive mode, Commission efforts certainly would not have achieved anything. As in ESPRIT major European corporations, including makers of telecoms equipment and large users of advanced telecoms services, played a crucial role in winning approval for collaborative initiatives. The Commission/industry alliance proved once again to be an effective starting point. In the story that follows the focus will be on collaborative R&D and the RACE program. But because the Commission advanced broad strategies for the entire telecoms sector (including standards and market opening), I will analyze those other dimensions of Commission strategy so as to place the R&D component in its context.

25. Catherine Le Bailly, "Fin du monopole aux Pays-Bas," *La Tribune*, 26 November 1985, p. 6; Laura Raun, "Signs of Vulnerability," *Financial Times*, 11 May 1988, Survey, p. 10; Terence Holsgrove and Marion Howard-Healy, *European Telecommunications at the Crossroads*, 7.

26. William Dawkins, "The Stakes Are High," *Financial Times*, 11 May 1988, Survey, p. 11.

Evolution of Commission Telecoms Strategy

As described in the previous chapter, Commissioner for Industry
Davignon brought to Brussels a keen appreciation of the challenges
facing Europe's telematics industries and the need for collaboration.
Already in 1979 Davignon was advocating the need for a European
industrial strategy to meet U.S. and Japanese competition in telematics. For Davignon telematics meant the broad sector comprising
components, computers, and telecommunications. IT (microelectronics and computers) received attention first, in the ESPRIT program. There were reasons for that ordering of priorities. For one,
the political problems in telecoms would be much trickier than in
IT, involving as they did the PTTs. An initial failure there would
put a damper on everything. A second reason for approaching telecoms slowly was that ESPRIT would cover many of the basic microelectronics technologies needed in telecoms systems. Thus, the
telecoms program could be a downstream, or applications, program, capitalizing on ESPRIT results.[27]

The first step, as mentioned in Chapter 7, came in September
1979. Davignon proposed that industry, governments, and the
Commission jointly work out an EC strategy for telematics. Regarding telecoms, he suggested that work focus on common standards, on an integrated digital network linking European institutions that would be a pilot for a future universal European digital
network, and on encouraging the growth of European data services.[28]

The following month Davignon sat with the general directors of
the telecommunications administrations of the member states as they
drafted a set of recommendations on EC telecommunications. The
Commission approved the recommendations, which were ratified
by the PTT general directors in December. The key points included:

1. CEPT should increase work on digital networks.[29]

2. Resources should be devoted to technical work leading to
harmonization.

27. The CEC makes this argument in its ESPRIT I proposal: *Proposal for a
Council Decision Adopting the First ESPRIT*, 11, n. 8.
28. *Agence Europe*, 24 March 1979, p. 10; 29 September 1979, p. 12.
29. CEPT is a body of twenty-six European PTTs. It has traditionally coordinated work on network interfaces and tariff arrangements.

3. New networks should evolve in coordination among the countries so as to provide similar capabilities.

4. Countries should harmonize the provision of new services, consulting CEPT when they plan new services so as to allow common guidelines.[30]

Of course, nothing the PTTs agreed to in CEPT was binding.

Meanwhile, the ITTF had drafted its document on telematics strategy, "La société européenne face aux technologies de l'information: pour une réponse communautaire," and submitted it to the Council in November 1979. Four of the paper's six points related to telecoms: create a European market for telecommunications and data-processing based on common standards; create a data-services industry competitive in world markets; create a communications network linking EC institutions and national governments; and develop common positions for world telecoms organizations.

The Council requested specific proposals in microelectronics and telecommunications, which the Commission submitted in July 1980. The paper on telecoms strategy argued that new services would be in demand after 1983 and they would require a harmonized, Europe-wide network. A draft recommendation requested the PTTs to open bidding for terminal equipment to all EC manufacturers and to set aside 10 percent of their network-equipment purchases for bidding from any EC maker. The Commission hoped for a Council decision by December 1980 so that the recommendations could be implemented over 1981–83.[31]

In the event, approval did not come until November 1984. The market-opening proposal obtained approval by not specifying the kind of European company (supplier or maker) it would benefit. The PTTs or other network operators would allow completely open bidding for all new terminals, and (beginning in 1986) for 10 percent of their annual purchases of network equipment and conventional terminals (like handsets, modems, telex).[32] The measures were weak. For one thing, they were passed in the form of a Recommendation (which is not legally enforceable). For another, the Council approved only an experimental two-year period for the open bid-

30. *Agence Europe,* 29 December 1979, p. 9.
31. *Agence Europe,* 18 July 1980, p. 6.
32. CEC, *Communication from the Commission to the Council on the Status of the Community Telecommunications Policy,* 5.

ding.[33] Consequently, as of autumn 1987, though the Germans had diligently listed their calls for tenders in the *Official Journal,* the Spanish and French had yet to do so.[34]

In June 1983 the Commission announced that it was preparing a proposal for Community action in telecoms. A document went before the Council at its Stuttgart meeting that month.[35] It declared that the goal was to stimulate the production and use of the most advanced equipment and services. The Commission was careful to stress that its increasing responsibilities in telecoms would in no way alter the duties of the PTTs nor affect national arrangements for the transfer of PTT revenues to other budgets. The Commission also recognized that the timing was propitious; the Ten needed to cooperate in order to ensure compatibility of new products and services during the period of rapid change due to digitization, satellites, fiber optics, and advances in microelectronics. Finally, the Commission requested the national governments to appoint representatives to a senior officials' group to discuss the establishment of a European telecoms body.[36]

Unfortunately for the Commission the ITTF was not always as tactful vis-à-vis the PTTs as the Commission had been in the Stuttgart document. Members of the ITTF, who were running the telecoms show within the Commission, spoke of creating a super-PTT at the EC level, to run the EC's networks. Naturally, this proposal enraged the PTTs. One of my informants, who had been a member of the ITTF at the time, said that it made the PTTs suspicious of Commission efforts and reluctant to participate.[37]

However, the idea of a senior officials' group did take hold. The Industry Council, meeting in November 1983, unanimously agreed with the Commission paper submitted the previous June in Stuttgart and asked the Commission to organize meetings with representatives of the industry ministers and the telecoms industries to prepare an action program by the end of the year.[38] Davignon and the ITTF responded by creating the Senior Officials Group for Telecommunications (SOGT), composed of a mix of representatives from

33. Ibid., 5.
34. Rene de Cazanove, "La quête d'un futur à douze," 53.
35. "Le CEE souhaite un 'marché commun' des télécommunications," *Les Echos,* 7 June 1983.
36. *Agence Europe,* 17 June 1983, p. 10.
37. Interview 50; de Cazanove, "La quête," 51.
38. *Agence Europe,* 7 November 1983, p. 7.

ministries of industry and economics and from the PTTs. The work of the SOGT provided the programmatic basis and initial political support for Commission telecoms initiatives.[39]

Meanwhile, at the end of September the Commission had proposed to the Council six lines of action in telecoms. The Commission's arguments in that early document touched on the issues that would remain central to telecoms discussions at the European level. First, European equipment makers needed "home-markets of sufficient size and, under European conditions, these can only be provided by the Community as a whole." The Commission pointed out that nine different switching systems had to try to survive on a fragmented European market, while three systems in Japan and four in the United States each benefited from a large home market. Second, the Commission stressed the threat from the United States and Japan, noting that Japanese industry was oriented by unified planning for the Information Network System and that Japanese exports in advanced telecoms gear were "already now advancing at an astonishing rate in major markets." Third, fragmented planning for ISDN and broadband networks both limited the scale of investment in future networks and services and "threaten[ed] the [competitiveness] of the European telematics industry."[40]

The six action lines were these:

1. "Setting medium- and long-term objectives at the Community level"

2. "Common action on research and development regarding the key segments of future development"

3. Development of interface standards so as to create an EC market

4. "Common development of the transnational part of the future telecommunications infrastructure in the Community," with emphasis on rapid development of advanced facilities for EC business users

5. Development of telecoms in less-favored regions

6. Opening of public procurement[41]

39. Interviews 28, 50; CEC, *Communication from the Commission to the Council on Telecommunications,* COM (84) 277 final, 2.

40. CEC, *Communication from the Commission to the Council on Telecommunications: Lines of Action,* COM (83) 573 final, passim.

41. Ibid., 8–13.

The first three objectives would be addressed in what became the EC's RACE program.

The newly constituted SOGT began a series of meetings with the Commission to discuss possible areas of cooperation. They met six times between November 1983 and March 1984; the consensus that emerged was the basis for Commission proposals for EC telecoms strategy in May 1984. The Commission again emphasized the dangers of U.S. and Japanese competition, stressing the competitive pressure unleashed by deregulation in the United States. Furthermore, the bottom line was that "the capacity to meet these challenges, and to cope in a timely manner with the opportunities born out of the development of telecommunications, is outside the capability of national operators on their own."[42]

The Commission placed a heavier emphasis than previously on the dynamic economic role played by telecoms. By 1990, the paper stated, telecoms would be the EC's largest economic sector, surpassing even the automobile industry, and accounting for 7 percent of Community GNP by the year 2000. It also noted that telecoms investment had a significant multiplier effect: For every ECU invested in telecoms infrastructure, there was a 1.5 ECU increase in total economic activity. In particular, demand was soaring among business users for advanced services. Thus, telecommunications would create new jobs and save jobs by increasing the competitiveness of firms. In fact, the document declared, telecoms "constitutes a special tool for reviving the economy and protecting employment in the Community."[43]

The action lines proposed in September 1983 became the outline, after approval by the SOGT, for an EC program in telecommunications. The main elements of the program were:

1. *Create a Community telecoms market via:*

    a. *Standards:* The Commission would identify standards requirements and set up a standards program that would (a) define priorities, (b) set up a timetable, (c) create procedures for cooperating with CEPT, and (d) establish means for monitoring and updating the standards work.

42. CEC, *Communication from the Commission to the Council on Telecommunications,* COM (84) 277 final, 2–3, 10.
    43. Ibid., 3–7.

b. *Type approvals:* The Commission would request the network operators (the PTTs and, in Britain, BT and Mercury) to implement mutual recognition of type approvals for terminal equipment, with assistance from CEPT.[44]

c. *Public procurement:* Network operators would be requested to invite bids from all EC makers for terminals and to open some minimum amount of their network-equipment purchases (10 percent) to bidding from all EC suppliers (one of the early recommendations, still held up by disputes over what was properly an EC supplier).

2. *Planning:* A group would be formed to analyze needs and coordinate introduction of new networks and services over the next twenty years, focusing on ISDN, mobile telephones, and broadband.

3. *Technology development:* The Commission was preparing, with the SOGT and industry, a program for R&D cooperation, which would be presented by the end of the year.

4. *Aid for telecommunications development in the less-favored regions.*[45]

In announcing the program, Davignon argued that the EC could not be complacent. "The situation in this sector is not bad but it is fragile. Up to the present this is the kind of situation in which the EC reacts the worst, by developing a sense of self-satisfaction," declared Davignon.[46]

The industry ministers in December 1984 approved the objectives proposed by Davignon. The third goal, development of the technologies needed for future networks and services, took shape in the form of RACE, to which the remainder of the chapter will be devoted.[47] Before analyzing RACE, however, I will examine the

44. *Type approval* refers to the process of verifying that a piece of equipment conforms to standards. The PTTs have historically set the standards and tested for compliance. Mutual recognition of type approvals would thus mean that a terminal approved in one national laboratory would be recognized by other countries as certified for attachment to the network.

45. CEC, *Communication from the Commission to the Council on Telecommunications,* COM (84) 277 final, 14–22.

46. *Europolitique,* 19 May 1984.

47. The fourth point, telecoms for the least-favored regions, became the Special Telecommunications Action for Regional Development (STAR) program. Budgeted at 780 MECU (to be matched by participating governments), the program was approved in 1987 to run for five years.

Commission's most recent synthesis of its telecommunications goals and strategies, the 1987 Green Paper.

The Green Paper

Telecommunications intersected with the 1992 project for unifying the internal market, and the result was the Commission's Green Paper. This lengthy document, circulated in May 1987 and released the next month, placed telecommunications reform in the context of revitalizing the European economy. In fact, the actual title of the paper is *Towards a Dynamic European Economy*. Its first paragraph declares: "A technically advanced, Europe-wide and low-cost telecommunications network will provide an essential infrastructure for improving the [competitiveness] of the European economy, achieving the Internal Market and strengthening Community cohesion—which constitute priority Community goals reaffirmed in the European Single Act."[48]

The Commission built its argument on the economic importance of advanced telecommunications, the technological changes sweeping the sector, and the reforms already underway in all the member states. The bottom line was that Europe would lose out, both as a competitive source of advanced telecoms equipment and services and as an economic player more broadly, if the changes were not managed jointly at the European level.

Backed by extensive documentation, the Commission concluded the Green Paper with ten positions intended to serve as guides for telecommunications development. I summarize them here.

1.   The PTTs may retain their monopoly on the provision and operation of the network infrastructure, though states may choose otherwise (as in the United Kingdom).

2.   The PTTs should continue to be the sole (or privileged, again in deference to the United Kingdom) providers of "a limited number of basic services." The Commission suggests voice telephony as the only clear candidate for the status of "basic services."

3.   Provision of all other services (including VANs) should be opened to private suppliers in competition with the PTTs.

48. CEC, *Towards a Dynamic European Economy*, 9.

4. For the sake of interoperability across the EC, the PTTs and other service providers must follow European and international standards.

5. The Commission, in consultation with the member states and telecoms administrations, should issue clear rules requiring the PTTs to grant fair access to the network to competing service providers.

6. The Community should benefit from a free market in terminal equipment, with the possible exception of the first telephone handset.

7. The operational and regulatory activities (type approval, interface specifications, tariff surveillance) of the PTTs should be split into separate administrations.

8. The Commission should monitor the commercial behavior of the PTTs to prevent cross-subsidization of their equipment and service activities to the detriment of competition.

9. The Commission should monitor private providers to avoid the "abuse of dominant position."

10. The Community's commercial policy (including competition rules) should apply to telecommunications.[49]

In sum, the Green Paper reasserted the Commission's goal to create open EC markets for network equipment, terminal equipment, and telecoms services. A key part of the Commission's approach was open-network provision (ONP), a transnational network operating on the basis of common standards and interfaces. With ONP service providers would have access to the network anywhere in the EC on the basis of established rules of usage and common tariff principles. ONP implied that customers would have access to advanced services from anywhere in the EC and that approved equipment could be attached in any country.

The Green Paper was intended to spark discussion within the Community, and it did. Private associations supported the Commission's plans. The International Telecommunications Users Group (INTUG), representing European national users' groups, strongly endorsed all the Green Paper proposals. The only reservations by INTUG were in areas where the group thought the Commission had

49. Ibid., Figure 13.

not gone far enough, such as in the preservation of the PTT monopoly on supplying the first telephone set.[50] The employers' federation and the Roundtable of European Industrialists also seconded the Commission's proposals.[51] In addition, the Commission carried on intensive discussions with the SOGT and with the national telecoms administrations.

As a result of all the feedback, the Commission prepared a second document, on implementing the Green Paper.[52] In it, the Commission reviewed the major points of the responses to the Green Paper. It identified areas of broad consensus, areas of general agreement but with criticisms (both from those who thought the proposals in question went too far and from those who argued that they did not go far enough), and areas to be further defined (like satellites). The Commission concluded that a strong consensus supported the major goals of the Green Paper and on that basis set forth a series of specific goals and deadlines, all to be achieved for the completion of the internal market in 1992. The timetable included opening the market for terminal equipment, open competition in services, ONP, common standards, separation of regulatory and operational responsibilities, and other supporting measures.

The Council supported all these objectives in a resolution passed in June 1988. The resolution, which was nonbinding, was without deadlines, but it did request the Commission to propose specific measures to achieve all the goals. In the remainder of this section, I describe the Commission's efforts to implement the Green Paper objectives, and the political rows they have provoked.

One of the central problems in creating a pan-European telecoms network was standards. Standards consist of technical specifications for how electronic information is to be packaged on the network and for how various pieces of network equipment are to talk to each other. Traditionally, the PTTs controlled the standards-setting process in each country and the procedures for verifying that specific pieces of hardware complied with the national standards.

50. George G. McKendrick, "The INTUG View on the EEC Green Paper," 325–29.

51. The Roundtable of European Industrialists is an association of about thirty of Europe's largest corporations from both EC and non-EC countries, including Philips, Siemens, Olivetti, Daimler Benz, Volvo, Fiat, Bosch, ASEA, and Ciba-Geigy.

52. CEC, Towards a Competitive Community-wide Telecommunications Market in 1992—Implementing the Green Paper on the Development of the Common Market for Telecommunications Services and Equipment—State of Discussions and Proposals by the Commission.

Common standards had been one of the Commission's priority themes. In fact, through the initiative of the Commission, in July 1984 the EC signed a memorandum of understanding with CEPT. Under this memorandum, CEPT produces common standards and specifications for equipment approval (type approval, in EC jargon) according to priorities established by the EC. The standards defined by CEPT are called Normes Européennes des Télécommunications (NETs). NETs are the basis for new services and equipment introduced or authorized by the PTTs.

The Commission chalked up another success when the Council, in July 1986, approved a binding Directive on mutual recognition of type approvals (Directive 86/361/EEC). The gist of the Directive was that once a specific type of terminal equipment has been certified as conforming to the relevant common standards, it must be accepted as approved in every other member country. The Directive employed the mutual-recognition approach to standards adopted by the Commission in conjunction with the 1992 program: Rather than suffer through the laborious process of elaborating EC standards, each country must accept as valid approvals granted in any other country. The legal effect is to make NETs mandatory for EC members. The Directive also assigned the Commission the task of preparing each year a list of priorities and a timetable for standardization and common technical specifications. The Commission would then transmit the list to CEPT, which would draft the appropriate NETs.

One priority area has been ISDN, the next step in network evolution. With ISDN, digitized networks will be able to carry simultaneously voice, data, text, and some images. For the range of new services to function across the EC, the ISDNs being created in each member country need to be built to common standards. The Commission proposed that the EC take actions to ensure pan-European ISDN. In the event, the Council passed a (nonbinding) recommendation "on the co-ordinated introduction of ISDN" in December 1986.

In the Green Paper the Commission proposed the creation of a permanent and independent European Telecommunications Standards Institute (ETSI). ETSI came into being in March 1988 despite some conflict over the overlap between its mission and that of CEPT in standards. However, CEPT had certain deficiencies—namely, a lack of permanent, specialized technical expertise, and a lack of representation for telecommunications industries. In ETSI, network operators (primarily the PTTs), equipment manufacturers, and telecoms users meet to establish technical specifications. Following Commission proposals ETSI produced a timetable for the coordinated introduction of ISDN services in Europe. The PTTs responded by signing a memorandum of understanding in April 1989 on the introduction of pan-European ISDN by 1992.[53]

The area in which the Commission most vigorously assaulted the old telecoms order was in opening to competition markets in terminal equipment and services. In both terminals and services the Commission sought to liberalize markets by issuing directives under Article 90 of the Treaty of Rome. The general principle of Article 90 is that enterprises to which states grant "special or exclusive rights" must conform to the competition rules of the Treaty. Most important, Article 90 charges the Commission to ensure application of its provisions and grants it the power to issue binding directives to member states. The key feature of such Commission directives is that they do not have to be approved by the European Parliament or by the Council of Ministers. The Commission used Article 90 against the PTTs, arguing that PTT monopolies constituted "abuse of dominant position" (Article 86), which restrains competition. Beginning in 1988 the Commission employed Article 90 to open EC markets in terminals and VANs. All member countries supported the goal of an open terminal market, but several opposed the Commission's unilateral tactic. In fact France (supported by three other countries) challenged the action in the European Court of Justice. The application of Article 90 in services was equally controversial, and several states objected to the goal as well as the tactic. A compromise was struck in December 1989, liberalizing VANs but leaving basic voice telephony and telex as PTT monopolies.[54]

In short the Commission spearheaded an effort that is creating a new telecommunications regime in Europe, one based on common standards and open markets.

THE POLITICS OF RACE

The various dimensions of the Commission's telecommunications efforts provide the context without which it would be difficult to

53. Hans Baur, "Telecommunications and the Unified European Market," 33–34.
54. See Wayne Sandholtz, "New Europe, New Telecommunications."

understand the role of the RACE program. RACE touches on several of the topics treated in the previous section, like standards and planning future networks and services.

Winning the Support of Industry and the PTTs

As with ESPRIT, the ITTF had prepared itself thoroughly for its assault on Europe's traditional telecoms structures. It had carried out its own assessments of the telecoms sector and had also commissioned a number of technical and economic studies from prominent consultancy firms, including Arthur D. Little International, Mackintosh International, McKinsey, and The Yankee Group. The outside studies provided ITTF members with the facts and figures they needed to be taken seriously. In other words, the studies were not a tool to convince the governments but rather a means of increasing the technical credibility of the Commission. When ITTF officials went before industry groups or the SOGT, they commanded reputable information.[55]

In fact the Commission began RACE with the same Roundtable of twelve companies that had designed ESPRIT. Why were the Twelve interested in the Commission's schemes for telecoms? The interest in collaborative R&D stemmed largely from the various aspects of technological change examined in Chapter 6. Innovation was dauntingly rapid, and it involved not only traditional telecoms technologies but also VLSI microelectronics, the technologies of data-processing and computer networking, and, with an eye to the future, optoelectronics. Collaboration was a way of covering more of the important technological bases than was otherwise possible. But one technology factor deserves special attention in the case of telecommunications: the rising costs of R&D and the attendant imperative for large markets.

The latest generations of public switches were extremely expensive to develop. The costs of developing digital exchanges ranged from $500 million to $1.4 billion. Future optical switches were sure to require substantially more resources. To amortize such large investments in R&D, companies had to achieve vast sales. The Commission cited studies showing that to be economically viable a switch would have to capture 8 percent of the world market. No national

55. Interviews 28 and 50.

market in Europe surpassed 6 percent of the world total.[56] Or, as
Rob van Tulder and Gerd Junne show, given development costs of
the magnitude cited and given that R&D consumes about 7 percent
of sales, a switch would have to capture a market worth $14–$16
billion. The difficulty was that no European country had a switch
market of that scale. The French market was worth about $10.9
billion, the German one $11.7 billion, and the British one $7.2 billion.[57]

Because of these pressures, the world telecommunications industry had been consolidating. Data show that telecommunications
companies were particularly active in the wave of interfirm alliances
that swept the telematics industries. Indeed, a list of interfirm alliances involving telecoms firms since 1980 would consume several
pages.[58] In addition, telecommunications manufacturers were merging at a striking rate. For its equipment business outside the United
States, AT&T entered into a joint venture with Philips. Plessey (United
Kingdom) bought the smaller company Stromberg-Carlson (United
States) and in 1988 formed a joint venture with GEC (United Kingdom), uniting their switching businesses. The biggest merger was
the 1986 purchase by CGE (France) of the overseas telecommunications subsidiaries of ITT, forming the world's second largest telecoms-equipment manufacturer, Alcatel. Ericsson (Sweden) subsequently bought the remaining French maker, CGCT. Siemens
established a joint venture with GTE (United States). The Italian
firm Italtel teamed up with AT&T, and Telettra swapped shares
with Spain's Telefonica. Consortia were forming in 1987 to compete for the pan-European mobile-communications market. Alcatel,
Nokia (Finland), and AEG (Germany) joined forces, as did Philips
and Bosch, and Ericsson had ties with Siemens, Matra (France), and
Orbitel (a joint venture of Racal and Plessey in the United Kingdom). By 1987 the telecommunications business was already a highly
concentrated one, with the ten largest firms accounting for 67 percent of world production.[59] The technological and market pressures
behind the trend of alliances and concentration made R&D collaboration attractive to European enterprises.

56. CEC, *Towards a Dynamic European Economy*, 90.
57. van Tulder and Junne, *European Multinationals*, 70.
58. See Pisano and Teece, *Collaborative Arrangements;* LAREA/CEREM, *Les
strategies d'accord;* and van Tulder and Junne, *European Multinationals*, 60–63.
59. Jacques Arlandis, "Le dilemme des quarante fabricants," 65.

The Commission had decided that starting anything in telecommunications would require action from two major groups: industry and the PTTs. Politics required that the Commission approach the PTTs first. The major companies active in telecoms equipment depended on close working relations with their PTTs and on PTT orders. For political reasons, therefore, they could not meet with the Commission until after the PTTs had done so. Shortly after the creation of the SOGT in late 1983, the Commission proposed meetings with the research arms of the PTTs (like the Centre National d'Etudes des Télécommunications in France or the Forschungsinstitut der Deutschen Bundespost in Germany). The telecoms administrations were extremely reluctant and ended up sending only low-level representatives.[60]

However, now the Commission could initiate meetings with industry. Representatives of the ITTF visited the leadership of the Roundtable companies, laid out the Commission's analysis of the challenges, and proposed that industry help in working up an R&D program aimed at broadband communications. The companies accepted. The result was the Planning Exercise in Telecommunications (PET). Over two and a half months in the summer of 1984 eighty experts from industry and the PTT research organizations met twice a week to determine the R&D activities that could benefit from Community-level cooperation. The nitty-gritty work took place in four technical panels organized by industry. In looking over the list of participants, I could identify only two national telecoms laboratories represented on the technical panels: Nederlandse PTT and the telecoms research center for the IRI-STET group.

The result of the PET was a volume of over 600 pages titled *Proposal for an Action Plan: RACE*. The document outlined a vision for broadband systems and proposed research areas and tasks to prepare for them. The overall objective, by consensus of the PTT labs and industry, was "Community-wide introduction of Integrated Broadband Communication (IBC) by 1995 taking into account the evolving . . . ISDNs." The IBC network envisioned here would eventually include all national and local networks and would subsume all the services presently offered on distinct networks, plus ISDN, cable television, and mobile communications. It would involve fiber-optic transmission down to and possibly including the

60. Interviews 28 and 50.

customer premises. IBC would unite in one network all business and residential services, from telephony to data transmission to mass-market HDTV. The proposed RACE program would provide the technological base for IBC.[61]

The body of the document detailed the R&D work to be carried out in RACE and proposed a separate program of IBC demonstrations and trials. The PET experts offered a timetable that included dates for the various phases of RACE, the demonstrators, and installation of the working system. They aimed for initial commercial availability by 1995. The RACE work would require an estimated 5,196 man-hours over five years. The R&D was divided into four areas:

1. Systems aspects, including systems architecture, subscriber environment, network subsystems, and operation

2. Requirements of users and service providers

3. Enabling and supporting technologies, including electronic components, optics and optoelectronics, and design tools

4. Communication software technologies.

A fifth area, not detailed in the report, would be low-cost, high-performance terminal equipment. For the four research areas, the document specified 115 separate tasks. For each task, objectives, technical approach, connections with other tasks, milestones, a time scale, and the resources needed (in man-years) were specified. The PET recommendations became the basis for the Commission's later proposals for the RACE program.

The Commission distributed the PET report to the PTTs for review. This was the hook that eventually brought in the telecoms administrations. The PTT directors saw that concrete, credible technology planning was going on with the heavy involvement of industry. It became clear, first, that if they were not involved, the PTTs would lose control of telecommunications development, and, second, that by participating the PTTs could inject their will into the proceedings. In fact one of my Commission informants told me that a PTT official vented her displeasure with events but conceded that the Commission had them "ensnared."[62] PTT participation be-

61. CEC, *Proposal for an Action Plan: RACE*, 0.1.1–0.1.4.
62. Interview 28.

came important later in the process, for although the decision to approve RACE was a political one and although various ministries (economics, industry, research) had input to the governments on the question, the PTTs were politically central.

Even though the PTTs were beginning to warm to the idea of RACE, the Commission ran into a temporary setback, largely of its own making. In late 1984 the Commission proposed to the SOGT a 1 BECU RACE program that was flatly rejected. The proposal was shot down so quickly that it never appeared in the press. Some in the ITTF had wanted to put a telecoms program on the books before Davignon left the Commission at the end of 1984. After that failure the Commission regrouped and drew up a smaller Definition Phase.

The Definition Phase

By March 1985 the Commission had put together a proposal for a preliminary, limited Definition Phase to precede the full-fledged RACE program. At the time the Commission foresaw a ten-year RACE program with a total EC contribution of about $400 million. Planners in Brussels hoped that the Definition Phase could be approved before the summer, so that it could issue a call for proposals in the autumn.[63]

With the proposal for the RACE Definition Phase, the Commission submitted to the Council a report on the R&D requirements in telecommunications. The report reiterated the importance of telecommunications to the European economies, especially as the convergence of telecommunications, data-processing, and audiovisual media proceeded. It also noted that R&D costs for telecoms equipment were soaring at the same time that the life expectancy of a public exchange was falling—ten to fifteen years for an analog exchange, less than ten years for a digital switch. A collaborative program would help ease some of the resulting pressures. The Commission argued further that the need for joint action was urgent given the efforts of Europe's competitors.[64]

The report also reaffirmed the consensus reached by the PTT research arms and industry in the PET—namely, that the objective

63. *Europolitique*, 16 March 1985.
64. CEC, *Report of the Commission to the Council on R&D Requirements in Telecommunications Technologies as Contribution to the Preparation of the R&D Programme: RACE*, 19–21.

should be an EC-wide transition to IBC. Toward that goal a collaborative R&D program should commence. The program's basic aims would include production of a reference model for IBC (its networks, terminals, and services), development of the technological foundation for that IBC, and close cooperation between RACE and existing telecoms standards organizations. The document also reproduced the timetable proposed in the PET—a Definition Phase (1985–86), a main Phase (1987–92), and a final phase (1992–97), with installation of the IBC network beginning in 1992 and operation in 1995.[65]

The proposal specified that the R&D work would be precompetitive (as in ESPRIT), executed via contracts between the Commission and EC organizations. The total EC contribution to the Definition Phase would be 22.1 MECU over eighteen months; the EC would cover half the costs of projects and participants would put up the other 50 percent. The proposal included a ceiling of 4.5 percent of the EC contribution that could be allocated to Commission staff.

The Definition Phase work plan encompassed two main parts, each subdivided into specific areas:

Part I: Development of an IBC reference model

  1. Development of an IBC network reference model

  2. Definition of the IBC terminal environment

  3. Future applications assessment

Part II: LLTRD

  1. High-speed ICs

  2. High-complexity ICs

  3. Integrated optoelectronics

  4. Broadband switching

  5. Passive optical components

  6. Components for high-bit-rate long-haul links

  7. Dedicated communications software

  8. Large-area flat-panel display technology

65. Ibid., 24–26.

For the three areas under Part I objectives, approach, methods, and tasks were specified. The areas under Part II also contained a number of specific tasks.[66]

The Commission's proposals for a Definition Phase were taken up by the Council in early June. The ministers of industry and of telecommunications, meeting jointly, did not arrive at any unanimous position on RACE. The three large countries expressed reservations about the level of funding proposed (22.1 MECU from the EC) and also argued that work on defining a reference model for IBC should be carried out in the association of PTTs, CEPT.[67]

The research ministers were more receptive than the industry and telecommunications ministers to the Definition Phase, although with reservations. The United Kingdom and Germany emphasized that a decision about the Definition Phase would not imply any commitment to a later Main Phase. Further, the research ministers stipulated that the Definition Phase work would comprise forecasts and estimates of R&D needs in the area, rather than actual R&D. Britain and Germany wanted the Definition Phase funding to come out of the existing Commission budget for R&D. The Commission ended up assenting to these conditions.[68]

The Research Council was able to reach a general political agreement on the RACE Definition Phase after a compromise was reached on the ties between the EC program and CEPT. The research ministers noted that the ministers responsible for telecommunications desired that CEPT play a central role in defining the IBC reference model. France in particular clung to that view, and Britain and Germany favored it as well. In essence, this was a move to ensure that the PTTs, through CEPT, would retain some control over planning for the future system. The director-general of the French PTT, Jacques Dondoux, speaking later in June at a meeting of CEPT, articulated that view: "A telecommunications Europe cannot be made only at the level of the EEC and industry; the network operators, who alone

66. CEC, *Proposal for a Council Decision on a Preparatory Action for a Community Research and Development Programme in the Field of Telecommunications Technologies: R&D in Advanced Communications-Technologies for Europe (RACE), Definition Phase*, passim.

67. *Agence Europe*, 5 June 1985, p. 5; Alain Bradfer, "Projet RACE: Les européens sont bien changeants," *Le Matin*, 5 June 1985.

68. *Agence Europe*, 5 June 1985, p. 5; Paul Cheeseright, "Setback for EEC Telecommunications Plan," *Financial Times*, 4 June 1985, p. 2; and Quentin Peel, "Broad Agreement on Ambitious EEC Plan for Communications," *Financial Times*, 5 June 1985, p. 2.

can ensure the interconnection of national networks, should play a greater role."[69]

The Commission wanted to retain control of the entire program, pointing out that CEPT included sixteen non-EC countries, that it had extremely tenuous relations with industry, and that it was organizationally and financially inadequate for the job. But facing the big three states, the Commission was forced to compromise, agreeing that the preparation of the IBC reference model would take place in close collaboration with the PTT association. The French research minister, Curien, declared himself satisfied that CEPT would make a strong contribution to the Definition Phase.[70]

The Research Council, however, could not pass a final decision because of a dispute involving a separate technology issue, the proposed EC laboratory for the handling of tritium, which is essential for research on nuclear fusion. The Italians withheld final approval of the Definition Phase in order to obtain approval for establishing the tritium laboratory at the Joint Research Center at Ispra in northern Italy. Only the French opposed the Ispra site. By the end of July the French had withdrawn their objection to the Ispra laboratory, and the Council of Ministers gave final approval to the RACE Definition Phase on 25 July 1985.[71]

As passed by the Council, the Definition Phase received a budget of 20 MECU, of which 14 MECU would go to the Part II projects open to bidding. The remaining funds were assigned to Part I (the IBC reference model). Part I was subject to restricted tendering because it was entrusted to CEPT and the PTTs. Incidentally, CEPT was not equipped to perform the reference-model work. In order to fulfill its role in RACE, CEPT created the first fixed structure in its history—a permanent secretariat, to be complemented by groups of experts hired for full time.[72]

The Council Decision authorizing the RACE Definition Phase defined operating procedures that had been settled in ESPRIT. Part II

69. Guillaume Goubert, "La longue marche des télécommunications vers la normalisation," *La Tribune,* 2 July 1985.

70. "RACE sur les rails," *La Tribune,* 6 June 1985; Alain Bradfer, "Projet RACE," *Le Matin,* 5 June 1985.

71. Philippe Lemaitre, "Les Dix confirment leur volonté," *Le Monde,* 6 June 1985; CEC, *Background Information for the Call for Tenders for the RACE Definition Phase,* 1.

72. Guillaume Goubert, "La longue marche . . . ," *La Tribune,* 2 July 1985.

projects had to involve at least two independent industrial partners from different member states. Participants would pay half the cost of each project, and the work would be carried out within the EC. The Commission would be responsible for managing the program in close cooperation with a Management Committee composed of two representatives from each member state. The Management Committee would have to approve any Part I contracts worth over 100,000 ECU and any Part II contracts over 400,000 ECU.[73]

The proposal and selection procedures mirrored those for ES-PRIT. Consortia of companies, universities, PTT laboratories, and other research organizations submitted proposals. Commission administrators, assisted by experts hired for that purpose, evaluated the proposals in several steps. First, they assessed the adequacy of the managerial arrangements and the fit of the technical content to the work plan. At this stage the proposals were anonymous. Next, the same set of evaluators looked at the qualifications of the proposers and their personnel. Finally, the evaluators examined the funding request. The proposals passed through two or three different sets of evaluators. The Commission then submitted a list to the Management Committee, recommending some proposals for quick approval, some for additional scrutiny, and some for rejection.[74]

After the research ministers had given general political approval to the Definition Phase, the Commission published in the *Official Journal* in April 1985 an "advance notice," inviting expressions of interest. Following final Council approval in July, the official call for tenders attracted more than eighty proposals involving 171 EC organizations. The Commission brought in about seventy experts from industry and academia for a period of two weeks to assess the proposals. In February 1986 the Commission announced the thirty projects that received funding; they would run until the end of 1986.[75] The Commission also contracted for special consulting studies on HDTV with several organizations. The broadband network would have the bit-rate capacity to carry HDTV signals. The HDTV studies reflected the emerging consensus that the extension

73. Council Decision of 25 July 1985, 85/372/EEC, *Official Journal*, no. L/210, 7 August 1985, pp. 24–26.

74. Interview 9; CEC, *Background Information for the Call for Tenders for the RACE Definition Phase*, passim.

75. Interview 9.

TABLE 8.2.   NATIONAL PARTICIPATION IN RACE
DEFINITION PHASE PROJECTS

	Overall Number of Projects	Number of Project Leaders
Belgium	8	2
Denmark	7	1
France	25	6
Germany	19	3
Greece	4	0
Ireland	3	0
Italy	16	3
Netherlands	14	1
Portugal	1	0
Spain	5	0
United Kingdom	26	14

SOURCE: Commission of the European Communities, *RACE: Consolidated Preliminary Report of the RDP Projects,* OTR 89 final (Brussels, June 1987), Annexes 2 and 3.
   Note: Subsidiaries of multinationals are counted in the country in which they operate.

of broadband networks into residences would be economically feasible only if they could offer entertainment services with potential for mass markets—like HDTV.

The largest of the Definition Phase projects was one to define a subscriber-premises reference model. It was headed by GEC of the United Kingdom, involved twenty-nine organizations, and received 3 MECU. The breakdown of national representation on the projects is presented in Table 8.2 European subsidiaries of five American-based multinationals figured among the winning consortia. In fact ITT affiliates were involved in thirteen of the thirty-one projects— more than any other company. Eight of the ITT participations were by Standard Elektrik Lorenz, the West German subsidiary. Other American companies represented in the Definition Phase included AT&T, GTE, Hewlett-Packard, and IBM. The participation of companies with U.S. ties disturbed some of the European firms, but neither the Commission nor the Council made an issue of it. The only stipulation in the rules governing the Definition Phase was that the R&D be performed within the EC.[76]

The RACE Definition Phase projects concluded at the end of 1986. Part I activities produced an initial IBC reference model via the efforts of the PTTs through the permanent nucleus of experts (the Groupe Spécial Large Bande) established within CEPT. An industry group advised the PTT-CEPT group working on the reference model. The model developed general outlines of the switching and transmission capabilities of the future IBC network and the kinds of services it could be expected to support. The services envisioned were grouped into four categories as shown in Table 8.3. The reference-model work also came up with three possible plans for evolution toward IBC. The first was based on circuit-switching techniques (in which connections are made by establishing a circuit that utilizes all the users' available channels). The second assumed asynchronous time division (in which a stream of "packetized" information of all kinds—voice, image, data—travels through a single channel, routed by the network switches). The third plan was based on the assumption of a fully optical network. Further clarification of the eventual services and research into the technological options would permit choices to be made among the plans.[77] The part of the Definition Phase open to tenders from EC companies, universities, and labs was also extremely successful.

Commission officials expressed a high degree of satisfaction with the Definition Phase and considered that it had legitimized RACE.[78] A survey of participants, including industry, universities, and network operators, revealed strongly positive reactions to the program. When asked whether the Definition Phase approach (exploring the technologies and the functional specifications simultaneously) was appropriate for European objectives and conditions, 85 percent answered affirmatively. The question "Do you consider that your participation in the [RACE Definition Phase] has

76. Richard L. Hudson, "Many in Europe Decry Inclusion of US Firms in EC Phone Project," *Wall Street Journal*, 13 February 1986; Richard L. Hudson, "EC Commission Gives 109 Grants on RACE Study," *Wall Street Journal*, 26 February 1986; Ivo Dawnay, "US Groups Prominent in Community Project," *Financial Times*, 26 February 1986, p. 2. By 1987 ITT's telecommunications subsidiaries in Europe had been purchased by France's CGE, forming Alcatel.

77. CEC, *RACE: Consolidated Preliminary Report of the RDP Projects*, 2-1–2-9; B. Catania, *The Many Ways Towards IBC*, 19–20.

78. Interviews 28 and 50.

TABLE 8.3.   IBC SERVICES CLASSIFICATION

Service Classes	Form	Broadband Services
Dialogue	Video	Videotelephony Videoconferencing
	Audio	Standard telephony High-fidelity conferencing
	Data	Bulk data transfer
Messaging	Video	Picture mail, video message
	Audio	User-to-user sound transfer
	Data	Mixed-mode messages (text, graphics, moving pictures)
Retrieval	Video	Videotex Home shopping/ banking Lending library
	Audio	High-quality sound library
	Data	Document retrieval, high-speed data retrieval
Distribution	Video	Broadcast and pay television High-quality corporate video HDTV
	Audio	High-quality broadcast and pay audio
	Data	Broadcast videotex Data distribution

SOURCE: Commission of the European Communities, *RACE: Consolidated Preliminary Report of the RDP Projects*, OTR 89 final (Brussels, June 1987), 1-2.

contributed significantly to a clarification of the issues and helped you in your own work?" elicited a positive response from almost 90 percent of respondents. Over 90 percent of respondents answered yes to the query "Has working directly together with your colleagues from other countries been of benefit and do you consider this to be important to the strengthening of Europe's posture?"[79]

The Main Phase

Before the Definition Phase had reached its conclusion, the Commission was already putting the final touches on its proposal for a Main Phase of RACE. Like ESPRIT II, RACE was packaged in the 1987–91 Framework Programme on R&D. RACE Main therefore became stuck in the political quagmire surrounding the Framework Programme, described in the previous chapter. Thus, although the Commission proposal went to the Council in late October 1986, it would be a year before the program was finally approved.

The Commission document proposing RACE Main rephrased slightly the overall goal that had been stated in the Definition Phase: "Introduction of . . . IBC taking into account the evolving ISDN and national introduction strategies, progressing to Community-wide services by 1995."[80] The change in the statement is the phrase "and national introduction strategies," reflecting increased sensitivity to the diverse plans and levels of advancement of the national telecoms administrations.

The proposal was for a five-year program with an EC contribution of 800 MECU, bringing the total program (with cost sharing) to 1,600 MECU and 10,000 man-years of work. The Commission raised the possibility of a second five-year phase but made clear that its present proposal in no way implied or required one. It argued that the Definition Phase had already demonstrated the need for a coordinated European approach to IBC development and that rapid follow-on work was needed to maintain momentum.

The Main Phase would include three principal parts:

Part I: IBC development and implementation strategies

  1. Development of common definitions of the IBC system and subsystems and a common approach to their introduction

  2. Further development of the IBC reference model

  3. Systems analysis and engineering to translate the reference model into functional specifications

79. CEC, *Proposal for a Council Regulation on a Community Action in the Field of Telecommunications Technologies: RACE*, 20–21.
  80. Ibid., 5.

Part II: IBC technologies

Collaborative R&D on key technologies for IBC equipment and services

Part III: Prenormative functional integration

Development of an open verification environment, which would permit the evaluation of concepts and experimental equipment, taking into account functional specifications and proposed standards emerging from Part I[81]

The Commission hoped for a launch date of 1 January 1987.

Because of the logjam holding up the Framework Programme, nothing happened on RACE through the first half of 1987. One Commission official considered the delay fortunate, as it allowed the ITTF (by then the Information Technologies and Telecommunications Task Force) time to digest the results of the Definition Phase and incorporate them into the first RACE work plan.[82] The draft work plan was finished in the spring and published in June 1987.

The work under Part I would focus on the continual updating and refinement of the reference model. In addition, it would include the mapping of the reference-model functions onto alternative implementation strategies. Each strategy would be broken down into a chronological sequence of frames, each frame representing a credible real situation. Each strategy was thus a model of the evolution, in concrete steps, toward IBC. The work plan contained two principal strategies.

Part II aimed at developing the technologies needed to implement the evolution strategies. The R&D therefore would pursue diverse technical options. The work in Part III would focus on two areas: the development of verification tools, and IBC applications pilot schemes (an addition). The pilot projects would ideally be transnational experiments to try out various IBC functions and equipment.

All three parts included a number of specific tasks, some ninety-nine in all, that proposals would be expected to address. Each task description included sections on background, objective, technical

81. Ibid., 28–30.
82. Interview 50.

approach, and key results and milestones. The work for Part I and the verification aspect of Part III would, as in the Definition Phase, center on the permanent broadband nucleus established in CEPT. The rest would be open to bidding.

As the Main Phase proposal sat on the shelf, it became increasingly clear that RACE had won broad support, both in industry and among the PTTs. The PTTs had discovered that they could exercise significant influence within RACE and had to maintain a role in the process that was underway. Indeed, it is striking that after the Definition Phase the PTTs did not oppose RACE. Commission officials noted that the problems in winning approval for RACE Main stemmed not from the national telecoms administrations but from the Framework Programme debates. One RACE administrator said that by the time he started circulating the draft work plan for the Main Phase, he had high-level contacts in the PTTs and was well received by them.[83]

Industry also lent strong support to the Commission's efforts in telecommunications. In October 1986, as the Commission was preparing its Main Phase proposal, the Roundtable of European Industrialists issued a paper on telecoms in Europe from the point of view of major business users. The Roundtable paper highlighted the problems caused for European businesses by diverse national equipment-approval rules, the lack of compatibility of services across borders, and the absence of coordinated planning for new services and networks. The paper summarized the Commission's initiatives (including coordinated ISDN, standardization, open markets, and RACE) and stated strong support for them: "Among users, the will exists to bring the EEC's objectives to fruition." The Roundtable recommendations paralleled the Commission's own.[84]

In January 1987 the industrial association Union des Industries de la Communauté Européenne (UNICE) submitted to the Commission a set of proposals for urgent action in the sector. It was the first public effort involving both makers and users of telecoms equipment to influence telecommunications policy at the EC level. The UNICE proposals paralleled the themes stressed by the Commission since 1983: an open European market for terminals, open public procurement, common standards, an end to PTT monopo-

83. Interviews 28 and 50.
84. Roundtable of European Industrialists, *Clearing the Lines: A Users' View on Business Communications in Europe*, 16–19.

lies, and continued development of EC R&D programs. The industrialists stressed the importance of precompetitive research.[85] All these themes emerged in the Commission's Green Paper, which circulated a few months later (as discussed previously in this chapter).

The Community telecoms industry enthusiastically backed the proposed RACE Main Phase. The only significant exception was Siemens, which, according to industry and EC sources, was reticent about the program.[86] Siemens executives explained that they were definitely in favor of strong European standards and that cooperative R&D was all right as long as it remained precompetitive. However, Siemens thought it was ahead of RACE objectives and would possess a competitive advantage in the future.[87] Even so, Siemens ended up participating in five of the first batch of Main Phase projects. In general, though, RACE had solid backing from industry.

The Commission had published in the *Official Journal* of 5 December 1986 a call for "expressions of interest" in RACE Main. The following March the ITTF distributed a book listing the names and addresses of the nearly 600 respondents (some organizations submitted multiple responses, in different areas).[88] In June the draft work plan was released, and on 1 July 1987 the Commission published in the *Official Journal* a "call for a reserve list of tenders." The call took the form of a "reserve list" because the Framework Programme—and with it RACE—had not been passed by the Council. But the Commission wanted to be able to move quickly once the anticipated approval finally came. By the October deadline, the Commission had in hand ninety-five proposals, involving 320 organizations.[89]

As detailed in the previous chapter, the member states finally reached a political agreement on the Framework Programme in July 1987, with formal ratification in late September. The component

85. Marc Paoloni, "Les P et T contre l'Europe?" *La Tribune,* 24 January 1987; William Dawkins, "Industry Outlines Telecoms Policy for EEC," *Financial Times,* 23 January 1987, p. 3.
86. Interviews 50 and 60; Philippe Lemaitre, "La Commission européenne veut consacrer 5,5 milliards de francs aux télécommunications," *Le Monde,* 21 October 1986.
87. Interview 45.
88. CEC, *RACE: Expressions of Interest.*
89. Interview 9; CEC, *RACE '88: The RACE Programme in 1988,* 1.

parts of the Framework Programme had to be approved by the Council individually; this approval came for RACE on 14 December 1987, with Community funding of 550 MECU (down from the 800 MECU originally hoped for by the Commission).[90]

The administration of the Main Phase follows the pattern set by ESPRIT and the Definition Phase. The RACE Management Committee must approve the yearly work plans and the list of projects chosen by the Commission. The prime contractor for each project must report monthly (in summary fashion) on technical and financial progress. In addition, each project is subject to an annual review, with the results to aid in updating the work plan and also possibly in revising the projects themselves. The Commission also provided for the project managers of all RACE projects to meet jointly two or three times a year to ensure communication among projects. The entire program was to be assessed after thirty months of operation and again at the completion of the five years. The Regulation establishing RACE left open the possibility of a second five-year phase.[91]

The September ratification allowed the Commission to proceed with the evaluation of the proposals and negotiation of contracts. The evaluation of the proposals followed the same criteria and procedures as in the Definition Phase and utilized the services of seventy-three outside experts.[92]

In the end forty-six projects were retained, with an EC contribution of 186 MECU over three years.[93] The first batch of projects covered most of the tasks in Parts I and II of the work plan. Work began in January 1988. As in ESPRIT II, countries of the EFTA were permitted to participate as full members of the consortia, though they could not receive EC funds. Of course, through Part I work managed by CEPT, all twenty-six CEPT members had an indirect hand in RACE.

With the first year's projects underway, the Commission began the first annual revision of the work plan. The process built on contributions from and consultations with some 2,000 parties. The

90. CEC, *RACE '88*, 1; CEC, *European Telecoms Fact Sheet 6: The RACE Programme*.
91. *Official Journal*, no. L/16, 21 January 1988, p. 35.
92. CEC, *Background Information for the "Call for a Reserve List of Tenders" for the RACE Programme: General Information*, 9–11.
93. CEC, *RACE Workplan '89*, v.

TABLE 8.4.   NATIONAL PARTICIPATION IN RACE
MAIN PHASE PROJECTS

	Overall Number of Projects	Number of Project Leaders
EC countries		
Belgium	23	6
Denmark	22	2
France	59	11
Germany	55	22
Greece	20	1
Ireland	13	4
Italy	35	7
Luxembourg	0	0
Netherlands	32	4
Portugal	15	2
Spain	38	1
United Kingdom	70	21
EFTA countries		
Austria	2	0
Finland	14	1
Iceland	0	0
Norway	10	0
Sweden	21	2
Switzerland	7	0

SOURCE: Commission of the European Communities,
RACE '89 (Brussels, March 1989), Annex A.
Note: Total number of projects: 84.

1989 work plan became the basis for a second call for proposals, which was issued in June 1988. The second round of proposals was to focus especially on Part III of the work plan, functional integration—the demonstration and testing of the technologies needed to support and integrate in a single network the services envisioned in the reference model. With the second round of contracts, eighty-four RACE projects were underway. Table 8.4 shows the levels of national participation in RACE through the second round. RACE

projects involve 362 organizations; of the 230 EC companies participating, 90 are small firms.[94]

Although no one can predict the full range of future IBC applications, they will probably be composed of a limited number of
basic functions, also called service primitives. The work plan for
1989 included tasks aimed at developing concepts and tools for verifying that emerging IBC components and subsystems (and eventually entire systems) perform the needed basic functions. To this
end a number of new tasks were added to Part III, including several
applications pilots that will test prototype IBC services for specific
classes of users. The areas to be addressed in the applications pilots
include:

1. Banking, finance, and insurance

2. Media and publishing, including high-quality image transfer

3. Manufacturing, including integration of design and production

4. Retail distribution and teleshopping

5. Medical-records transmission, including image-based rec-
ords

6. Transport and traffic management

7. Needs of people with disabilities

Work on the applications pilots will interact with the other parts
of the program to refine the model and the technologies.[95]

CONCLUSION

Given the situation in European telecommunications circa 1980,
few could have predicted the extent of collaboration that emerged
over the next eight years. At the beginning of the decade, national
PTTs held a monopoly not only on telecommunications networks,
equipment, and services, but also on policy-making. By 1988 that
was no longer the case. Because of Commission leadership, EC states

94. CEC, *RACE '89*, 1–3.
95. CEC, *RACE Workplan '89*, passim; and CEC, *Background Information for
a "2nd Call for Proposals" for the RACE Programme: General Information*, passim.

now pursue joint actions in planning future networks and in R&D. They have taken important collaborative steps in standards, type approval, and the opening of markets for equipment and services. In effect the Commission is at the center of an emerging new regime for telecommunications in Europe; it forced CEPT to assume new roles and created ETSI.

The theoretical framework of this study proves useful in explaining the cooperation. First, on the demand side, purely national strategies of the traditional European kind were discredited. Technological change and deregulation abroad were undermining the PTT monopolies, and governments increasingly realized that the nationalistic fragmentation of the past would not serve their present and future needs. Thus, policy-makers were in the adaptive mode, searching for new approaches to the telecommunications sector; major studies and reforms were in progress in all the EC countries by the mid-1980s.

Second, political leadership was crucial in organizing cooperation. The perceived inadequacy of national strategies and the adaptation undertaken by governments created an opportunity for policy leadership on the part of the Commission. It seized the initiative and convinced the governments and the PTTs that cooperation was essential. The same transnational industrial coalition that brought forth ESPRIT allied itself with the Commission and lobbied powerfully for telecommunications collaboration.

Finally, RACE benefited from organizational arrangements that provided for a satisfactory distribution of the benefits. The program is big enough to offer a large number of pieces of the action. Additionally, as in ESPRIT, states participate in a modified version of à la carte: They contribute to the whole program but participate only in parts that their domestic organizations find attractive.

For all the debate and confusion surrounding telecoms liberalization in the United States and Japan, the process in those countries was simple compared with what transpired in Europe. European states not only began to liberalize but also acted to end decades of nationalistic fragmentation in telecommunications. The results constitute a significant political achievement.

NINE

EUREKA

An Intergovernmental Counterpoint

In the spring of 1985 French President Mitterrand proposed that the Western European countries band together to promote their high-technology industries. In an uncommon display of speed, eighteen European countries had agreed on the general goals and outline of the EUREKA program by the following July. Despite earnest efforts by the Commission to include EUREKA in the EC fold, the program received an independent Secretariat, located in Brussels but not tied to EC structures. EUREKA is a strange hybrid. Its objective is to encourage cross-national consortia for the development of marketable products based on advanced technologies. But project approval and funding depend entirely on national governments. EUREKA, then, provides an instance of international collaboration under close national supervision. Rather than taking shape around existing IOs, EUREKA came into being as an ad hoc, intergovernmental bargain.

The immediate trigger for Mitterrand's proposal was the American invitation for European participation in SDI research. But it would be inaccurate to portray EUREKA as a European response to Star Wars. Rather, EUREKA took shape in a context of grave European doubts as to the viability of Europe's high-technology industries. Furthermore, EUREKA came into being just as the EC telematics programs were moving into high gear. Indeed, ESPRIT was becoming the object of Euro-enthusiasm; it was widely seen as an example of what Europe could and should accomplish. Preparations for RACE were gathering speed. In addition, the Ariane launcher was starting to fill its order books, and Airbus was finally cracking into world markets. In other words, EUREKA came into being in a context that had two crucial elements: an intense policy crisis centered on European telematics; and recent European successes that raised the value of collaboration. Given these two powerful aspects of the setting, SDI appears clearly as little more than a motive or trigger.

The EEC programs ESPRIT and RACE play an even greater role in the story of EUREKA than is apparent on the surface. I will argue that governments reacted to the success of the Commission's telematics programs by trying to preclude an even greater role for the Commission in European technology policy. In other words, the European governments were convinced that collaboration was a vital element in their technology strategies, but that did not necessarily mean they wanted ever greater powers transferred to the EC. European governments wanted to exercise tighter direct control over the collaboration than they did in ESPRIT and RACE.

EUREKA also provides an instance of technological collaboration that is nearer the market than the other telematics programs. While ESPRIT and RACE are restricted in principle to precompetitive R&D, EUREKA is explicitly aimed at developing high-technology products that have good chances of success on world markets. That EUREKA has this goal reinforces my argument that proximity to the market is not the most important obstacle to technological collaboration. Competitive commercial interests do, of course, have to be accommodated. But the range of organizational arrangements for assuaging competitive fears is greater than conventional wisdom usually acknowledges.

ORIGINS OF EUREKA

The central fact about SDI was that Europeans saw it not only as a security policy but also as an industrial policy. Naturally, the implications of SDI for defense and the Atlantic alliance provoked vigorous debate in Europe. My discussion will leave to one side the thorny military issues surrounding the notion of SDI. Instead, I will focus on the industrial-policy implications of Star Wars as these were perceived in Europe and as they influenced European technological collaboration. The bottom line was that SDI constituted a threat to the high-technology industries of the European allies. Table 9.1 provides a chronology of the major events in the evolution of EUREKA.

Of primordial importance is the context into which the Star Wars proposal was launched. European governments and industrial elites were in the midst of a profound high-tech policy crisis. In this setting President Ronald Reagan, in March 1983, first mooted the idea of SDI. It quickly became apparent that regardless of whether it produced anything like a shield against intercontinental ballistic missiles, SDI would fund vast amounts of research in technologies that had important commercial applications. By 1985 SDI had both a budget and technical content. The total expenditure for five years was projected at $26 billion; the budget for 1985 was $1.4 billion.[1] The R&D projects would be grouped in five main areas: surveillance, directed-energy weapons, kinetic-energy weapons, systems analysis, and supporting missions.[2] Even without knowing the precise content of the projects, one could guess the technologies that would have to be developed from the tasks SDI aimed to achieve. The work on laser, x-ray, and particle beams could have extensive applications in biomedical research and manufacturing. The detection and tracking technologies, including sensors of various kinds (optical, laser, radar, infrared), could have applications from transport to factory automation.

The greatest industrial advances would almost certainly come from the technologies required to coordinate and command the various parts of a space-based defense system. The key areas were supercomputers, software, and communications. The task of processing, at unprecedented speeds, the thousands of sensory inputs from satellites and electronic intelligence posts would demand qualitative and quantitative leaps in computer technology. Similarly, the precise aiming and coordinated firing of the various weapons would make unprecedented demands on computers, requiring AI techniques that do not exist at present. Linking the various ground- and space-based parts of the system would require reliable, high-capacity, secure communications links. All of this coordination depended on high-speed, highly integrated semiconductors. Star Wars research addressed the next generation of computer components—

1. Stowsky, *Beating Our Plowshares into Double-Edged Swords,* 2.
2. Council on Economic Priorities, *Star Wars: The Economic Fallout,* 11–15.

TABLE 9.1.   EUREKA CHRONOLOGY

Year	Event
1982	EEC: In December the ESPRIT pilot phase is approved by the Council.
1983	United States: In March President Reagan announces his vision of SDI.
	France: In September the French issue a memorandum on European technological collaboration.
1984	EEC: In February ESPRIT Phase I is approved by the Council.
1985	United States: In March Weinberger sends a letter to U.S. allies inviting participation in SDI research.
	France: In April Mitterrand proposes EUREKA to European partners.
	EEC: In June the Milan Summit is held; the Commission proposes the European Technology Community.
	EUREKA: In July the first EUREKA ministerial conference is convened in Paris.
	EEC: In July the RACE Definition Phase is approved by the Council.
	EUREKA: In November the second EUREKA ministerial conference is convened in Hanover; the first ten projects are announced.
	United Kingdom: In December the British sign an agreement on SDI research.
1986	Germany: In March the Germans sign an agreement on SDI research.
	United Kingdom: In March the British government agrees to allow state funding for EUREKA projects.
	EUREKA: In June the third EUREKA ministerial conference is convened in London; it approves new projects and concludes a formal agreement on a secretariat.
	EUREKA: In December the fourth EUREKA ministerial conference is convened in Stockholm; new projects are announced.
1987	EEC: In June the Commission publishes the Green Paper on telecoms.
	EEC: In July the new Framework Programme is approved after over one year of delay.
	Germany: In July the Germans announce DM 500 million to be spent on EUREKA projects.
	EUREKA: In September the fifth EUREKA ministerial conference is convened in Madrid; new projects are announced.

TABLE 9.1.   (CONTINUED)

Year	Event
	EEC: In December the RACE Main Phase receives formal approval.
1988	EUREKA: In June the sixth EUREKA ministerial conference is convened in Copenhagen; new projects are announced.
1989	EUREKA: In June the seventh EUREKA ministerial conference is convened in Vienna; new projects are announced.

optoelectronics—which will operate on the basis of pulses of light instead of electricity.[3] These were precisely the areas about which Europeans were intensely concerned in the early 1980s.

That Star Wars had immense industrial possibilities was obvious to the Europeans, but the initial public debates chiefly addressed the military wisdom of the initiative. In time attention began to focus on the economic impact of Star Wars research. As early as 1984 American observers were pointing out that the Department of Defense was virtually running an industrial policy for the United States. Economist Robert Reich compared the Pentagon to Japan's MITI: "The technology that MITI has targeted is identical to what the Pentagon is targeting: new materials, lasers, software, robotics and supercomputers."[4]

As American public debate shifted from an exclusive focus on the strategic merits of the program to broad consideration of its economic impact, SDI proponents began to play up the commercial and industrial benefits SDI would bring to the United States. SDI planners already in 1984 had created the Innovative Science and Technology Office with the goal of encouraging the interchange of research results between academia and small businesses.[5] In the fall of 1985 Lieutenant General James Abrahamson, director of the

3. See Stewart Nozette, "A Giant Step Forward in Technology," *New York Times,* 8 December 1985, p. F2; "Will Star Wars Reward or Retard Science," *Economist,* 7 September 1985, pp. 95–96; Malcolm W. Browne, "The Star Wars Spinoff."

4. Quoted in Michael Schrage, "Defense Budget Pushes Agenda in High-Tech R&D," *Washington Post,* 12 August 1984, p. F1.

5. Wayne Sandholtz, Jay Stowsky, and Steven K. Vogel, "The Dilemmas of Technological Competition in Comparative Perspective: Is It Guns v. Butter?", 8.

Strategic Defense Initiative Organization, which oversees the SDI program, created the Office of Education and Civil Applications, later renamed the Office of Technology Applications. Its purpose was "to find commercial product and process applications for SDI" innovations.[6] Indeed, in the second half of 1985, as EUREKA was moving from a political consensus to a concrete program, U.S. administration and SDI officials shifted their emphasis from SDI as a defensive shield to SDI as a research program with multiple payoffs.[7] By October 1985 General Abrahamson was testifying to Congress that he was "determined that we not miss the opportunity to capitalize on the results of S.D.I. research and apply it across all facets of our economy and society," and that he aimed to stimulate "the widest possible use of S.D.I.-related technologies, consistent with security considerations, for civil use."[8]

Europeans focused on the technological and industrial consequences of SDI, especially after the March 1985 invitation of American Defense Secretary Caspar Weinberger to U.S. allies to participate in SDI research. In a letter to all the NATO governments, as well as those of Japan, Australia, and Israel, Weinberger invited research participation in SDI by scientists and companies from those countries.[9] Europeans bristled at the sixty-day deadline for responses that Weinberger maladroitly included in the invitation, though the prospect of tapping into SDI technologies appealed to some.

The British, for instance, moved from a dismissive position on strategic grounds to a formal agreement to participate in SDI research. Foreign Secretary Sir Geoffrey Howe in March 1985 referred to the plan as "a new Maginot Line of the 20th century, liable to be outflanked by relatively simpler and demonstrably cheaper countermeasures."[10] Yet before long the British recognized that SDI would be a greenhouse for technological advances, both military and civilian. In April 1985 Defense Minister Michael Heseltine, acknowledging growing European interest in the program, said that

6. Rosy Nimroody, William Hartung, and Paul Grenier, *Star Wars Spinoffs: Blueprint for a High-Tech America?*, 9.
7. Judith Miller, "Allies in West Lend Support to 'Star Wars,' " *New York Times,* 30 December 1985, p. 1.
8. Quoted in Browne, "Star Wars Spinoff," 24, and Miller, "Allies in West Lend Support to 'Star Wars,' " 1.
9. "U.S. Asks Allies to Join Space-Weapons Effort," *New York Times,* 27 March 1985, p. A12.
10. Quoted in Marino de Medici, "Europe Sees SDI as Two-Edged Sword," 10.

the United Kingdom was motivated by the potential technological benefits.[11] The Thatcher government, while cautious about actual deployment of a strategic defense shield, began negotiating a research role in SDI. By October 1985 Heseltine could declare that the obstacles to British participation in SDI research had been overcome, including assurances from the United States that British firms would fully share the technologies developed.[12]

Thus, the United Kingdom became the first country to sign a formal government-to-government accord with the United States regarding Star Wars participation. Significantly, the Americans never assented to the British demand that the agreement guarantee to British firms contracts worth at least $1.5 billion. Nevertheless, Heseltine expressed confidence that British scientists and industries would benefit from "real opportunities, real contracts and real jobs." The British also pointed out that the accord did not represent complete U.K. endorsement of the planned defensive system, recalling the agreement between President Reagan and Prime Minister Thatcher that deployment would not proceed without negotiation with the Soviet Union.[13]

The Germans were also extremely cautious in their official pronouncements on SDI. The German government consistently stopped short of endorsing the military goals represented by Star Wars, while pointing out the strategic dangers—decoupling, new arms races, violation of the Antiballistic Missile Treaty. Still, Chancellor Kohl did endorse the SDI research program in principle in the spring of 1985. At the same time German concerns about the technological and industrial effects of Star Wars rose. Lothar Spath, premier of Baden-Württemberg, was quoted as urging European participation in SDI research because of its importance for nonmilitary, high-technology competition.[14] Chancellor Kohl expressed a similar view: "SDI will give the U.S. a big technological advantage whether or not the research leads to its intended goals. Highly industrialized countries like West Germany and the other European allies must not be left

11. Paul Lewis, "Briton Sees Allied Interest in 'Star Wars' Research," *New York Times,* 3 April 1985, p. A10.

12. James M. Markham, "Britain in Accord with U.S. on Full Role in 'Star Wars,' " *New York Times,* 31 October 1985, p. A12.

13. Joseph Lelyveld, "Britain Signs Pact on Research Role in U.S. 'Star Wars,' " *New York Times,* 7 December 1985, p. 1.

14. Barnaby J. Feder, "Europe Is Split on 'Star Wars,' " *New York Times,* 2 May 1985, p. 32.

behind."[15] The federal government made it clear that German participation in SDI research depended on U.S. assurances of its willingness to share the technologies developed.[16]

Yet the Germans did not rush into a formal agreement regarding the research. According to officials in Bonn, the British agreement in December 1985 would ease the way for Germany to negotiate a similar arrangement. The German government decided that same month to negotiate such an accord, over the strong opposition of Foreign Minister Genscher and his Free Democratic Party. The negotiations with the United States proceeded, and Germany signed an accord in March 1986. Significantly, and underscoring the German interest in the technological and industrial potential of the program, the Kohl government sent Economics Minister Martin Bangemann to sign the agreement in Washington, D.C.[17]

The German and British accords on SDI do not commit their governments to contribute in any way to the program. In fact all they do is formally declare that companies from their countries may contract with the Strategic Defense Initiative Organization or American firms for Star Wars research. But companies from nations that have no such agreement, like France, can also participate in SDI projects. In short, the agreements appear to be largely symbolic, demonstrating British and German endorsement of the research but not necessarily of the full operational deployment of such a system. More important, for the purposes of my argument, the agreements reflected the deep-seated European insecurities of the period over high technology. The agreements embodied the European fear that large American defense R&D programs posed an industrial threat to Europe in areas where it was already weak.

POLITICS OF EUREKA

Of the European countries, France most decisively rejected both the strategic premises of SDI and formal participation in the program. The month after Weinberger's letter to the allies, French President Mitterrand proposed a European high-technology research program, which took shape as EUREKA. From the earliest stages EU-

15. Philip M. Bofley, "Allies Back 'Star Wars' Studies, Not Deployment," *New York Times*, 10 February 1985, p. 14.
16. John Tagliabue, "Europeans Exploring Potential Role in 'Star Wars' Research," *New York Times*, 29 March 1985, p. A9.
17. "Bonn to Get 'Star Wars' Role," *New York Times*, 27 March 1986, p. A16.

REKA was touted as a civilian program, whose aim would be to promote the development of commercial products. Nevertheless, French security concerns hovered just below the surface. In fact, it was initially unclear whether the French were proposing a military or a civilian R&D program, especially because the areas suggested by France for cooperative R&D closely paralleled those covered by SDI. I will show that both military and industrial concerns motivated the French. EUREKA therefore illustrates the difficulty of untangling military and economic interests in advanced technologies. The root of the problem is that the same leading technological sectors are crucial to both economic growth and modern weapons systems.

French Initiative

Like the British and Germans, the French moved from an initial focus on the military implications of SDI to a consideration of its technological and industrial potential as well. Unlike Germany and the United Kingdom, France remained adamantly opposed to the idea of space-based strategic defenses (French security interests are discussed in the next section). President Mitterrand however became acutely aware of the technological boost SDI threatened to give to American industries. After the Weinberger invitation the French president recognized that SDI might draw away European scientific and technological talent. This fear was the basis for his EUREKA proposal.

On 17 April 1985, barely three weeks after the Weinberger invitation, French Foreign Minister Roland Dumas presented a letter to his EC colleagues announcing that President Mitterrand proposed to the other EC countries the "immediate creation of a technological Europe, aiming to permit our continent to master all of the high technologies."[18] The letter stated that recent collaborative successes like Airbus, Ariane, ESPRIT, and the EC project for joint research on nuclear fusion proved that Europe could cooperate effectively if the means were adequate. Thus, even though it eventually took shape outside of EC structures, EUREKA benefited from the collaborative enthusiasm generated in part by EC programs. In fact the French suggested that the program could employ the fund-

18. "Au Conseil des Ministres: le projet 'EUREKA' d'une Europe de la technologie," *Le Monde*, 18 April 1985, p. 30.

ing model developed in ESPRIT, with governments and industry each contributing half. The Dumas letter also said that France and Germany together were proposing the program. For political reasons that I will explain later, the initiative had to come from France. The proposal was for a European Research Coordination Agency (whence EUREKA), with a light administrative structure, to coordinate research in optoelectronics, new materials, high-power lasers, particle beams, large computers, AI, and high-speed, high-integration semiconductors.[19]

This close duplication of SDI technological domains and some initial French vagueness about whether EUREKA was intended to be a military or a civilian program led many to conclude that it was a direct response to Star Wars and the precursor to a European strategic defense system. Such a conclusion is not accurate. It is worth recalling that Mitterrand had instituted a rapid turnaround in French technology policies in 1983. After two years of national-champion technology policies and go-it-alone economic policies generally, France suddenly became a supporter of European cooperation. In the fall of 1983 Mitterrand circulated his memorandum to the EC governments calling for a "European industrial space," including increased technological collaboration to meet the U.S. and Japanese threat. Furthermore, France became a vigorous supporter of the Commission's telematics programs, ESPRIT and RACE. The EUREKA proposal fits in this pattern.

Furthermore, the planning for EUREKA took place among ministries and agencies charged with civilian technology programs. Military planners apparently had no role. In fact several participants from different positions in the French administration confirmed that the idea for EUREKA was particularly "Elyséenne"—that is, the initiative came from President Mitterrand and a handful of his advisors at the Palais de l'Elysée. Mitterrand, according to officials at the Research Ministry and at the Elysée, had expressed the desire to launch a major European initiative; he wanted to be seen as doing something for Europe.[20]

The Elysée had reflected on the American SDI program and other defense R&D programs (like the Strategic Computing Initiative). According to one member of Mitterrand's staff, the consensus was that although American military R&D programs were probably not

19. "Au Conseil des Ministres," Le Monde, 18 April 1985, p. 30; Agence Europe, 25 April 1985, p. 7.
20. Interviews 10, 14, and 39.

as effective as Japanese civilian ones in promoting commercial interests, nevertheless the U.S. programs were virtually certain to lead
to dramatic technological advances. The Elysée thought this development was potentially dangerous, as it could leave France dependent on the United States for certain advanced technologies. Recent
difficulties in obtaining specialized parts for the French (civilian)
earth-observation satellite because of American military restrictions
had served as a reminder of French technological vulnerability.[21]

The notion of a European technological program came in large
part from Jacques Attali, one of Mitterrand's close advisors. By the
end of 1984 Mitterrand and his circle were entertaining the idea.
Discussions broadened to include representatives of the Research
Ministry and the Foreign Ministry. Research Minister Curien, a scientist with extensive European experience as president of the European Science Foundation and former chairman of the Council of
the ESA, supported the idea provided that it not appear too French
but rather be clearly European.[22] Discussions revealed a general
consensus that precompetitive research (as in ESPRIT, which was
regarded as a success) was not then the primary problem. Rather,
downstream and applied activities at the European level were seriously lacking. As a result, a new market-oriented program would
have to be outside of EC structures, which were unsuited for commercial activities. In addition French officials had learned from various committee meetings and conversations that there was industrial support for such a market-oriented collaborative program.

Into this setting arrived the Weinberger letter. The invitation for
European scientists and corporations to join in SDI research posed
several new challenges. The one that preoccupied Mitterrand, in
addition to the technological advances that would benefit American
industries, was that European scientists and engineers would be attracted to SDI's forefront research and massive funding. Such a brain
drain would divert already scarce research talent and facilities des-
perately needed by European industries for commercial competitiveness. An analysis of the EUREKA proposal in *Le Monde* depicted the prevailing attitude in the French government as being
that Europe would participate in SDI as a subcontractor, leading
to fragmentation of European technological resources and brain
drain.[23] Furthermore, European companies would be unlikely to

21. Interview 39.
22. Interview 14.
23. "EUREKA," *Le Monde,* 19 April 1985, p. 1. See also de Medici, "Europe

benefit from commercial spinoffs because American security controls would prevent them from having full and timely access to the results.

French officials articulated these concerns repeatedly in the months following their April EUREKA proposal. At a meeting of foreign and defense ministers of the Western European Union (WEU)[24] French Foreign Minister Dumas declared, "If we do not rapidly coordinate our policies, nothing will prevent our researchers, our capital and our industries from giving in to the temptation of immediate ad hoc collaboration [in SDI], with the role of the Europeans being reduced to one of a sub-contractor."[25] Weeks later, in ruling out official French participation in SDI research, President Mitterrand employed similar language.[26]

Mitterrand's position was backed by a study conducted by the Centre d'Analyse et Prévision, with the help of public establishments (like the CNES and the Commissariat à l'Energie Atomique) and industry (like Matra and Thomson). The study concluded that although SDI would contribute to American industrial competitiveness in some areas, participation by European firms in SDI was not the best response, as commercial spinoffs were uncertain and European participation would be sought in areas where Europe was ahead anyway. The study also concluded that EUREKA would better meet European goals, as it would address Europe's greatest need, commercial development of technologies.[27] In early June Mitterrand stated that SDI and EUREKA did not compete in their objectives, as one was military and one was civilian. Rather, they com-

---

Sees SDI as Two-Edged Sword," 10: "The French government is preoccupied by the possibility that the research and development of SDI may work as a powerful 'vacuum cleaner' by attracting the 'best and the brightest' and the most promising work out of the European laboratories and industries."

24. The WEU is a forum for the discussion of European security issues; the members are the Benelux countries plus France, Germany, Italy, and the United Kingdom.

25. Henri de Bresson, "M. Dumas: 'le defi pour l'Europe est d'abord technologique,' " *Le Monde*, 25 April 1985, p. 2.

26. Bernard Brigouleix, "Pourquoi la France ne participera à l'IDS," *Le Monde*, 7 May 1985, p. 4. Although France rejected a government-to-government agreement on SDI research, by late 1985 the government was saying that it would permit French companies, even the state-owned ones, to participate in SDI projects on their own initiative; Judith Miller, "Allies in West Lend Support to 'Star Wars,' " *New York Times*, 30 December 1985, p. 1.

27. Philippe Lemaître, "Le programme EUREKA doit proposer un champ d'applications civiles plus large que le projet strategique de M. Reagan," *Le Monde*, 8 May 1985, p. 4.

peted for budgets and brain power, and with EUREKA both would remain in Europe.[28]

Thus the Weinberger letter acted as a "catalyst," according to French Research Minister Curien, for the launching of ideas that France had been formulating for many months.[29] The first step was to contact the Germans. German support would be essential in any case, and Mitterrand had been diligently cultivating the Franco-German relationship as the foundation of Europe. Foreign Minister Dumas alerted his German counterpart, Genscher, that the French government was discussing internally the possibility of a European high-technology program. According to officials at the BMFT, Genscher had been thinking along similar lines (he was a staunch opponent of SDI participation). But because he did not feel that he had enough political backing in the CSU/CDU-FDP coalition to propose such a program, Genscher asked Dumas to take the initiative. However, both French and German officials agree that the EUREKA initiative was intended to be European from the beginning, that it was not meant to be a Franco-German project with other countries tacked on as needed.[30] As I will show later, the Paris-Bonn connection was also important in giving EUREKA concrete form and substance.

French Security Interests

The foregoing discussion makes it clear that the principal motives for EUREKA, as well as its bureaucratic origins and development, were based on civilian technology concerns. But French military concerns also played a role, albeit indirect and below the surface. SDI posed a fundamental challenge to France's traditional reliance on an independent nuclear deterrent. If strategic defenses really made nuclear weapons obsolete (or even less effective), the foundation of French security policy would be undermined. Hence the French position that SDI would be strategically destabilizing. Not surprisingly, SDI triggered French thinking about advanced military technologies.

An initial indication that at least some members of the French defense community were thinking about SDI-type technologies came

28. Jacques Isnard, "M. Mitterrand au salon du Bourget: l'IDS et Eurêka ne sont pas des projets concurrents," *Le Monde,* 1 June 1985, p. 24.
29. Quoted in Dickson, "France Seeks Joint European Research," 694.
30. Interviews 10, 11, 39, and 61.

in an article by a French general writing under a pseudonym. The general argued that France should lead a European effort to develop its own space-based missile system.[31] In fact French defense officials began talking about an anti-tactical-ballistic-missile defense, which would require the same sorts of technologies as Star Wars.[32] Another tack was taken by a Defense Ministry study that concluded that a first step could be the improvement of French nuclear weapons, enabling them to penetrate new defenses.[33] A final event of interest is that Defense Minister Charles Hernù created in June 1985 a group for space studies within his office. Its mission was to analyze satellite technologies for surveillance, listening, and telecommunications, which were needed to control French forces around the world. Said Hernù: "We need to prepare our presence in space."[34] In short, the defense community in France was definitely pondering the implications of new missile defenses and space-based defense systems.

Given French military concerns, it is not surprising that the initial EUREKA proposal should be somewhat vague as to its civilian or military orientation. In fact the letter sent by Dumas to his EC counterparts underscored the "numerous spinoffs, in all civil sectors, of course, but also in the military domain" of a European R&D program.[35] Additional ambiguity stemmed from the fact that the technologies proposed by the French as important for collaboration matched SDI technologies, as we have seen. Furthermore, the Dumas letter tried to make a subtle distinction between space-based weapons and "military functions with peaceful ends . . . that consist of listening, seeing and communicating."[36] No wonder some of the small countries were suspicious of the proposal as not being clearly enough civilian.

The problem was that a military program could not have been sold to the European partners. Germany, for instance, faced with

31. John Vinocur, " 'Star Wars' Plan for Europe Urged," New York Times, 8 March 1985, p. A11.
32. See Robert C. Toth, "Europeans Seeking a 'Mini-Star Wars' System," Los Angeles Times, 19 December 1985, p. I24.
33. Richard Bernstein, "France Seeks Arm to Evade Space Radar," International Herald Tribune, 13 November 1985, p. 1.
34. Jacques Isnard, "M. Hernù crée un état-major de l'espace," Le Monde, 5 June 1985, p. 1.
35. "Les Sept en quête d'une réponse commune à l'initiative du Président Reagan," Le Monde, 23 April 1985, p. 4.
36. "L'initiative française vise à créer une structure souple 'associée' à la CEE, mais comportant d'autres partenaires," Le Monde, 20 April 1985, p. 2.

directly competing programs, would probably have had to go with the Americans. It is virtually certain that Britain would have made the same decision. One observer commented in *Le Monde* that the prevailing attitude in Paris seemed to be that in the absence of a European defense community the best course was to emphasize civilian R&D because the technologies could find military applications at a later stage.[37] In fact, France could ensure that much of the technological base would be in place via civilian programs because the technologies were the same. This argument seems to have entered the thinking of Dumas, who at the WEU meeting in April stated, "The challenge for Europe is first of all technological; the military challenge will come later, perhaps in a form that cannot be foreseen."[38] One French commentator later noted that it was important to recognize a single technological base for both SDI and EUREKA, and argued that Europe had to master that base.[39] Finally, it seems significant that among the first French companies to put forward specific EUREKA projects were two major defense contractors in which the state held a stake: Matra and Aerospatiale.[40]

In short, though the military angle did not enter explicitly into EUREKA politics, French concerns for defense technologies were at work below the surface. The evidence I have cited is circumstantial but strongly suggests that the French decided that because the technologies were the same, a civilian program could cover many of the right bases for later military applications.

Selling EUREKA

When the French floated the EUREKA idea in mid-April 1985, it was as half-baked as President Reagan's SDI proposal had been two years previously. There had been no technical studies, no attempts to include industry in the process, and no advance discussion with potential European partners. Once the proposal was in the open, the task of giving it substance and selling it to the other countries began.

37. "EUREKA," *Le Monde*, 19 April 1985, p. 1.
38. Henri de Bresson, "M. Dumas: 'le defi pour l'Europe,' " *Le Monde*, 25 April 1985, p. 2.
39. Jacques Isnard, "Des industriels partagés entre EUREKA et IDS," *Le Monde*, 9–10 June 1985, pp. 1, 14.
40. David Marsh, "EUREKA Emerges from Its Incubation Period with Wide European Support," *Financial Times*, 16 July 1985, p. 2.

Initial public reactions from other European countries were generally favorable, though cautious because the proposals completely lacked details. The week after the Mitterrand proposal, Foreign Minister Dumas attended the meeting of the WEU and lobbied for EUREKA. On the agenda at the WEU meeting was discussion of a unified European response to the Weinberger invitation, which had been advocated by Jacques Delors, president of the Commission, and by the West Germans. Although the WEU failed to approve a joint response to SDI, it did give tentative endorsement to the idea behind the French proposal.[41] Over the following months, French delegations went on the stump for EUREKA, selling the program in Europe's capitals. The delegations were led by Claude Arnaud, an advisor to the Foreign Ministry, and involved representatives from the Industry and Research Ministries, as well as some of Mitterrand's top advisors. Research Minister Curien made numerous visits himself.[42]

The first delegation traveled to Bonn immediately after the Dumas letter circulated. The initial reception there, according to a French participant, was "very cold"[43] despite the strong support of Foreign Minister Genscher, who had played a behind-the-scenes role in getting the proposal launched. In fact, on the occasion of the WEU meeting Genscher expressed himself as "very much in favor" of EUREKA.[44] The Germans were particularly opposed to any sort of centralized, bureaucratic agency, and the French proposal had vaguely used the term *agency*.[45] Discussions between the French and the Germans, nevertheless, continued over the following months, with the objective of reaching a common position on EUREKA in advance of the European summit to be held in June.[46]

German leaders were trying to decide whether EUREKA conflicted with SDI participation, which Kohl was inclined to pursue. As it became clear that EUREKA was not necessarily an alternative to SDI but rather parallel to it, the German government began to favor the European program. By the end of May Foreign Minister

41. Quentin Peel, "Delors Wants EEC Framework for EUREKA Project," *Financial Times*, 30 April 1985, p. 2; *Agence Europe*, 25 April 1985, p. 4.
42. Interviews 14 and 39.
43. Interview 39; *Agence Europe*, 24 June 1985, p. 5.
44. *Agence Europe*, 24 April 1985, p. 9.
45. Interviews 39 and 61.
46. David Housego, "Britain and France Back High-Tech Plan," *Financial Times*, 22 May 1985, p. 1.

Genscher could announce general approval of the program, while other government spokespersons stated that EUREKA did not rule out participation in SDI.[47] By the end of June the German government indicated clear backing for EUREKA, after German companies had begun to express an interest in the program. French and German foreign and research ministers reached agreement in principle on a definition of EUREKA to be submitted to the Milan Summit.[48]

The British too were unenthusiastic at first about Mitterrand's proposal. The same member of the French team who described the first German reaction as "very cold" said that the British reaction was "glacial." One British official declared that the only firm that could carry out international collaboration in Europe was IBM. According to one of the United Kingdom's point men in EUREKA, British reticence was due to the need to find out what the program was. The proposal came virtually without warning and took them by surprise. But once EUREKA began to take form, the British became strong supporters. A key British official declared that the United Kingdom seized on EUREKA so quickly and contributed so much to its incarnation that the French were surprised. An advisor in Mitterrand's office confirmed that the United Kingdom had worked hard to give EUREKA shape and definition.[49]

After a visit by French Research Minister Curien to London and two other senior French missions, Foreign Secretary Howe met with his French counterpart, Dumas, and clarified that Britain was not reticent about EUREKA. Howe also expressed the British view that U.K. participation in SDI was not incompatible with the European program.[50] Within two weeks Howe had written to Dumas to confirm that Great Britain would "fully participate" in EUREKA. Another British spokesperson said that the U.K. government shared the French view that the object of EUREKA was above all to close the technology gap between the United States and Europe.[51]

47. Steven V. Roberts, "Senate Vote Backs Test of Antisatellite Weapon," *New York Times,* 26 May 1985, p. 7; "Le chancelier Kohl devrait donner son accord," *Le Monde,* 29 May 1985, p. 44.

48. Rupert Cornwell, "W. Germany Backs EUREKA High Technology Project," *Financial Times,* 27 June 1985, p. 2; *Agence Europe,* 28 June 1985, p. 12.

49. Interviews 39 and 47.

50. David Housego, "Britain and France Back High-Tech Plan," *Financial Times,* 22 May 1985, p. 1.

51. "Londres confirme son intention de 'participer pleinement' au programme EUREKA," *Le Monde,* 4 June 1985, p. 48.

Other countries expressed varying degrees of support for the initiative. Italian Foreign Minister Giulio Andreotti stated that the EUREKA proposal merited further study and noted that Italy had always supported European research collaboration as a response to the technology gap separating Europe from the United States. By the end of May Andreotti was saying that "the EUREKA project is a boat which the Europeans must catch sooner or later, but preferably sooner." After a summit meeting with French President Mitterrand, Italian Prime Minister Bettino Craxi affirmed that Italy would cooperate in the EUREKA program, while not ruling out Italian participation in SDI.[52] The Netherlands and Belgium came out in favor of the French proposal, though with reservations about research cooperation conducted outside the EC framework.[53] Ireland, Denmark, and Greece were wary of the military implications of some of the technological areas suggested by France. The smaller states in general feared that the larger ones would dominate the program and monopolize the benefits.[54]

Nonstate Actors

I have argued that IOs can exercise policy leadership during crises. Here, the EUREKA case provides a counterpoise to ESPRIT and RACE in this sense: Whereas in ESPRIT and RACE the Commission played a primordial role, in EUREKA states consciously sidestepped EC institutions. Yet the national governments sought an R&D program outside the EC precisely because the Commission had been so successful in seizing the policy initiative with the previous two programs. ESPRIT and RACE marked an unprecedented degree of Commission involvement in industrially significant technology policies.

The previous cases have shown that industry plays a crucial role in international technology collaboration. Because industry performs the major share of high-technology R&D (again, as opposed to basic science), industrial support is necessary for an international R&D program to succeed. In EUREKA, though a transnational co-

52. *Agence Europe,* 24 April 1985, p. 9; 24 May 1985, p. 3; "Italian Premier Backs France's EUREKA Plan," *New York Times,* 15 June 1985, p. 7.
53. *Agence Europe,* 24 April 1985, p. 9; 28 May 1985, p. 10.
54. Quentin Peel, "Delors Wants EEC Framework . . . ," *Financial Times,* 30 April 1985, p. 2; Quentin Peel, "Community Support Grows for EUREKA Programme," *Financial Times,* 5 June 1985, p. 2.

alition did not play the primary role that the Roundtable companies did in ESPRIT and RACE, success depended in the end on the participation of firms. Indeed, EUREKA would have no content other than the projects proposed by cross-national industrial groupings.

The Commission, like everybody else, was taken by surprise with the French proposal for EUREKA. Commission President Delors welcomed the proposal as a counterbalance to SDI and immediately declared that EUREKA should be brought within the EC framework. Delors argued that the new program should be part of the EC even if all the member states were not involved. He promised to propose at the next EC summit (at Milan in June) institutional reforms that would allow such a program to proceed without unanimity of the Ten. Funding could be shared by the EC and national governments, or it could derive purely from national contributions, as in Euratom.[55] Over the succeeding months, the Commission lobbied vigorously to take a central role in administering EUREKA.

In mid-May Delors announced that the Commission was readying a proposal for a "Community of Technology." The Commission presented its memorandum on a European Technology Community in late June to the EC heads of state in Milan. The Commission proposed that the Technological Community receive 6 to 8 percent of the EC budget by 1990, pointing out that the member states had expressed a desire to increase the R&D expenditure of the EC faster than the overall budget. Commission Vice-President Narjes argued that in view of the challenge from the United States and Japan, the EC had to enhance dramatically its R&D role.[56]

The Commission's memorandum listed a number of technologies that could be included in such a program. The list closely paralleled the areas suggested for EUREKA: IT, biomolecular engineering, new materials, lasers and optics, large scientific instruments, broadband communications, new means of transport, the utilization of space, marine sciences, and education and training for technology. EUREKA would be subsumed in the EC's broader program. The organizational form of the Technology Community was left open, though Commission officials mentioned the possibility of a new treaty.[57]

55. Quentin Peel, "Delors Wants EEC Framework . . . ," *Financial Times,* 30 April 1985, p. 2; *Agence Europe,* 30 April 1985, p. 6.
56. *Agence Europe,* 24 June 1985, pp. 5–6.
57. *Agence Europe,* 24 June 1985, p. 5.

The Commission continued to argue for a central role in EU-REKA. But because France, Germany, and the United Kingdom were adamantly opposed to making it an EC program, and to any sort of strong bureaucratic structure, the Commission's hopes were frustrated. The Commission became just one participant in the EU-REKA discussion.

Industry support became vital to EUREKA, even though it was not involved in the early stages of planning. In fact French officials said that before Mitterrand's April 1985 proposal, there had been no specific consultations with industry about such a program. Even though the Dumas letter spoke of an "agency," it quickly became apparent that any organizational structure for the program would be skeletal and that the main initiative for making it work would come from industry. French participants in early EUREKA planning asserted that they wanted the program to be oriented toward the development of marketable products and that no planners, either at the EC or national level, could tell industry how to fit market demands. Therefore, industry would propose the kinds of projects to be included.[58]

The preferred term for EUREKA's format became *bottom-up co-operation*. The bottom-up approach was assured as it quickly became clear that both Britain and Germany were opposed to any sort of bureaucracy and wanted industry to drive the program.[59] Just before the first EUREKA ministerial conference, in July 1985, French Research Minister Curien declared that the program should be rooted firmly in industry and aim at the development of commercial products.[60]

So the task of drawing industry into the EUREKA discussion became a priority. The French took the direct route. Within weeks of the EUREKA announcement the presidents of the nationalized companies were called to a breakfast meeting and urged by the minister of industry to support President Mitterrand's initiative by quickly submitting proposals.[61] In September 1985, at another breakfast meeting, Curien and Minister for Industrial Redevelopment Edith Cresson urged industrialists to come forward with projects.[62] French

58. Interviews 10, 11 and 14.
59. Interviews 47 and 61.
60. David Marsh, "France Wants EUREKA Role for Industrialists," *Financial Times*, 15 July 1985, p. 30.
61. According to an executive from one such firm; Interview 8.
62. *Agence Europe*, 12 September 1985, p. 13.

projects were among the first to receive the EUREKA label, and French participation has remained high.

German and British companies were initially skeptical, especially because the program had no structure, no plan, and no money.[63] Enterprises in both countries came around in time, as the program acquired definite contours. In July the Bundesverband der Deutschen Industrie, the major German industry association, expressed support for all initiatives that promoted European technological collaboration. The group considered EUREKA a logical continuation of existing programs and hoped that it could be executed without excessive bureaucracy and preferably within EC structures.[64] British firms also ended up supporting EUREKA. In fact, by October 1985 prominent British companies (including GEC, Plessey, and Thorn-EMI) were praising the program but roundly criticizing the government for its refusal to offer partial state funding for EUREKA projects.[65] Most significantly, national EUREKA administrators in France, Germany, and the United Kingdom all reported that some companies were participating in EUREKA projects without any state aid.[66] In short, the essential industrial support was forthcoming.

EUREKA TAKES SHAPE

For almost a year after the French first proposed it, EUREKA slowly acquired organizational form and technical content. One of the first steps was to expand the discussions out from the EC countries. Of course, Spain and Portugal, set to join the EC, were included. The Dumas letter had also suggested that the program should have links with the technologically advanced non-EC states like Sweden, Norway, Austria, and Switzerland.

At the Milan Summit in late June the heads of the EC states agreed on an ad hoc committee to address specific plans for EUREKA. The first meeting would be convened by France on 17 July 1985 and would include delegations from the Ten and from Spain, Portugal, Sweden, Norway, Austria, and Switzerland. In addition the Council and the Commission would be represented on the committee. The inclusion of the EC organs came at the insistence of the

63. Interviews 5, 46 and 61.
64. *Agence Europe*, 27 July 1985, p. 10.
65. Peter Marsh, "Anger over Lack of State Cash for EUREKA," *Financial Times*, 11 October 1985, p. 8.
66. Interviews 30, 59 and 61.

Benelux countries. Interestingly, Delors did not associate himself with the Benelux proposal, though he explicitly hoped EUREKA would be placed within the EC context.[67]

Evolution of the Program

When the first EUREKA conference convened in Paris on 17 July, seventeen countries (Finland had since joined the club) and the EC were at the table. The countries were represented by foreign and research ministers; the EC, by Delors, Narjes (now Commissioner for Industry), and two other senior officials. The result of the conference was a brief communiqué that affirmed unanimous support for the notion that Europe had to unite "its energies and abilities in the field of high technology." The communiqué declared that the participating countries considered EUREKA established as of that date, even though it was silent about organization, technical content, and funding mechanisms. These "details" would be addressed at a second ministerial conference to be convened by Germany before 15 November. A group of senior civil servants was formed to work out concrete propositions.[68] The remainder of this section addresses each of these three issues in turn, after looking at the role of the EEC.

EUREKA and the EEC   The chief organizational issues in EUREKA were what kind of administrative structure to give it and whether it should be formally tied to the EC. The debate over EC affiliation boiled down to the large countries against the small. At the July conference that declared EUREKA in existence the Benelux countries expressed their fear that in an intergovernmental arrangement like that proposed for EUREKA the larger countries would dominate the program. Italy sided with the Benelux countries in calling for a major role for the Commission. At the insistence of this alliance, the Commission and the Council were formally represented in the high-level group charged with preparing specific proposals for the November conference. The Council of Foreign Ministers also requested the Commission to study ways of coor-

67. *Agence Europe,* 3 July 1985, p. 5; 4 July 1985, p. 6.
68. David Marsh, "Mitterrand Urges Europe to Support EUREKA Project," *Financial Times,* 18 July 1985, p. 48; David Marsh, "French Bid to Control EUREKA Fails," 19 July 1985, p. 42; *Agence Europe,* 19 July 1985, p. 7.

dinating EUREKA with its proposed Technology Community, again reflecting the concerns of Italy and the Benelux countries.[69]

The three large countries, however, were adamantly opposed to fitting EUREKA within EC structures. France, Germany, and the United Kingdom were all satisfied on the whole with how the Commission had administered the ESPRIT program. If anything, the Commission had performed better than expected. But the big three also thought the Commission less well suited to run a program that aimed at commercial developments, and they wanted to avoid having too much technology-policy power flow to the Commission. Several government officials from the three large states acknowledged that EUREKA was intended to be a collaborative R&D program over which the governments could exercise direct control.[70]

Given the overwhelming political power that the big three wield in European affairs, their opposition torpedoed any central role for the Commission in EUREKA. One incident illustrates the political divide. An informal meeting of EC research ministers was scheduled for October to discuss the Commission's Technology Community proposals; the small states also wished to reach a common position at that meeting in advance of the EUREKA conference. The research ministers of France, Germany, and the United Kingdom did not even attend. The delegates who attended the meeting from the big three opposed any Community consultation about EUREKA prior to the ministerial conference scheduled for November and wanted to reach agreement on some major projects that could be included in EUREKA before consulting with the Commission.[71]

For its part the Commission submitted a communication on the Technology Community to the Council in late September. By this time the Commission recognized the political impossibility of bringing EUREKA into its purview. The Commission position was that EUREKA was complementary to its programs and, being closer to the market, could be seen as an extension of them. The Technology Community had three main parts: a comprehensive new Framework Programme, the overall EC plan for R&D, which could be expanded to cover the technical areas discussed in EUREKA; the goal of completing the internal market as laid out in the famous 1985 White Paper; and an emphasis on increased cooperation among

69. *Agence Europe*, 19 July 1985, p. 7; 29 July 1985, p. 5.
70. Interviews 10, 11, 15, 30, 47, 51, and 61.
71. *Agence Europe*, 21 October 1985, p. 10; 25 October 1985, p. 10.

European industrial enterprises. The Commission called for an increase in the level of EC resources devoted to science and technology to 8 percent of the budget by 1990.[72]

In the end the Technology Community proposal was left gathering dust on the shelf. By February 1986 the Commission was stating that EUREKA could take projects undertaken at the precompetitive stage in EC programs and could move them forward toward the market.[73] By the end of the year the Commission had resigned itself to a secondary role. In fact, in November 1986 the Commission sent to the Council a document, *EUREKA and the European Technology Community,* which I summarize in the next several paragraphs. In it the Commission reduced the Technology Community to the Framework Programme for science and technology. As shown in Chapter 7, the Framework Programme budget for 1987–91 was eventually pared from 10.3 BECU to just over 5.5 BECU.

The Commission acknowledged that the objectives of and the technologies included in EUREKA were similar to those of its own programs. Both aim at increasing the competitiveness of European high-tech industries. The COST program is the EC program most similar to EUREKA in that it involves non-EC countries, and states participate only in those projects that they choose. But COST covers precompetitive R&D. Furthermore, COST proposals originate with governments, and the work is conducted in national laboratories and institutes. In EUREKA the focus is on the market; companies propose the projects and carry out the work. The Commission also noted that EUREKA did not benefit from any overall strategic plan, as EC programs do.

The Commission recognized that EUREKA could complement its programs by carrying the work closer to the market. Therefore, the Commission stated, the programs must be coordinated to minimize overlap and take advantage of synergies. In areas where EUREKA and EC programs have similar interests, the Commission could coordinate the two by ensuring that information and appropriate results flowed between them. The document contained an extensive list of EUREKA projects matched with specific EC projects (many of them from ESPRIT and RACE) that address relevant topics. Other possible EC roles included pushing for the completion of the inter-

---

72. *Agence Europe,* 13 September 1985, p. 7; 27 September 1985, p. 5.
73. *Agence Europe,* 3 February 1986, p. 13.

nal market, especially for high-technology products, and actual participation by the Commission in some projects. The Research Council (then up to twelve members after the accession of Spain and Portugal) agreed with this view of the Commission's place.[74]

Organizing EUREKA   In the meantime, preparatory work for the November 1985 ministerial conference aimed at defining a structure for the program. Meetings in the fall included representatives of France, West Germany, the Council presidency, and the Commission. This working group prepared draft statutes to govern the program. Yet consultations with the eighteen participating countries revealed a lack of consensus on the Secretariat and on funding. The small states (led by Benelux and Italy) favored a strong central administration that could ensure that the large states did not prevent access by the smaller ones to projects and technologies. But France, Germany, and especially the United Kingdom feared bureaucratic immobility and wanted only a small office to coordinate the program. Disagreements were such that the ministerial conference convened at Hanover on 5 November approved a general Declaration of Principles and left details on the Secretariat and funding to be worked out by experts in early 1986.[75]

One thing that was clear at Hanover was that the large EC countries and the non-EC participants were strongly opposed to any major Commission role in the Secretariat. The response of the Commission was that the EC should continue to pursue the Technology Community as agreed at the Milan Summit. Commissioner Narjes argued that the Technology Community was especially important to protect the small states and that it could be coordinated with EUREKA. Through these arguments the Commission was trying to accommodate itself to the attempt to bypass it.[76]

The Declaration of Principles affirmed that EUREKA aimed at "closer cooperation among enterprises and research institutes" in order to develop "products, processes and services having worldwide market potential and based on advanced technologies." The document also made clear that the "projects will serve civilian pur-

74. CEC, *EUREKA and the European Technology Community*, passim; *Agence Europe*, 10 December 1986, p. 7.

75. Rupert Cornwell, "Defining EUREKA's Contours," *Financial Times*, 4 November 1985, p. 2; *Agence Europe*, 27 August 1985, p. 4; 4 November 1985, p. 8.

76. *Agence Europe*, 8 November 1985, p. 7; 9 November 1985, p. 8.

poses." Projects had to include participants from more than one European country. Parties to a EUREKA project would decide on the form of cooperation and arrange for adequate financing.[77]

The process for launching a EUREKA project is as follows. Each country appoints a High Representative to coordinate EUREKA work in her country. Once firms or institutes have agreed on a project, each one approaches the EUREKA coordinator in its home country. Each coordinator verifies compliance with EUREKA objectives and criteria and informs the other national coordinators of potential projects. Once all the participants have earned the EUREKA approval of their home governments, the High Representatives meet as a group to inform the EUREKA Council of Ministers (consisting of representatives of the national governments and the Commission) of the approved projects. The Council of Ministers, the highest executive body, therefore merely announces the projects that have already been approved. The High Representatives also prepare the meetings of the Council of Ministers, which has a rotating chair.[78]

Finally, the Declaration called for the establishment of "a small and flexible EUREKA secretariat." Its functions would be to collect and disseminate information on the projects, to help in finding partners for interested organizations, and to provide support for the meetings of the Council of Ministers and the High Representatives. The Declaration concluded by noting that EUREKA was not meant to supplant any existing programs of technological cooperation but rather to supplement or extend them.[79]

The High Representatives of the eighteen countries (Turkey had joined) met in January 1986 and agreed on a Secretariat of six members, including a Commission representative. The Secretariat would run a computerized database of the projects and interested parties.[80] The location remained a minor problem. France and Germany had suggested Strasbourg, but the other states were reluctant, as that location would give the program an even stronger Franco-German appearance than it already had.[81] Belgium later put forward Brussels as the logical choice, given the presence there of other major European institutions. A high-level EUREKA meeting in May

77. EUREKA, "Declaration of Principles Relating to EUREKA," 1–2, 4.
78. Ibid., 4–5.
79. Ibid., 6–7.
80. *Agence Europe,* 25 January 1986, p. 11.
81. *Agence Europe,* 9 November 1985, p. 8; 23 November 1985, p. 3a; 25 November 1985, p. 3.

reached tentative agreement on Brussels, though the French maintained a reservation. Compromise was possible if the head of the
Secretariat were French. This was the arrangement that received
formal approval at the ministerial conference in London of June
1986. In the end, the Secretariat comprised seven administrators
(plus a small secretarial staff), headed by the Frenchman Xavier
Fels. The four largest EC countries and the Commission would each
contribute 13.7 percent of the operating costs of the Secretariat,
with the remainder shared by the other countries. The Commission
also agreed to provide the Eurokom electronic information system
developed in ESPRIT.[82] Thus, it was nearly a year after the initial
agreement in Paris establishing EUREKA before the participants could
agree on the organizational structure for the program. Still, a number of specific projects were already underway.

Technical Content   Initial meetings to work out the technologies that might be covered by EUREKA began shortly after France
sprung the proposal in April 1985. Bilateral meetings of experts
took place the same month, between France and Great Britain as
well as between France and West Germany.[83] The technical interests of the three large countries in the end defined the program's
technical content. But that is not saying much since by its nature
EUREKA would include anything two European parties wanted to
do together relating to advanced technologies. France, Germany,
and the United Kingdom held differing preferences for the sorts of
technologies the program should address. These differences did not
produce major political battles because each country was allowed
to bring what it pleased to this technological potluck. Still, a brief
look at the diverse technical emphases will illustrate the varying
interests accommodated by EUREKA.

The most fundamental difference in approaches to EUREKA's
technical content was that separating Germany from the other two
big countries. According to participants from each country, Germany from the beginning was interested primarily in research having to do with environmental protection and, secondarily, with infrastructure (for example, transport).[84] As the Hanover conference

82. *Agence Europe,* 8 March 1986, p. 4; 12 May 1986, p. 13; 26 June 1986,
p. 8; 2 July 1986, p. 7.
83. "M. Curien: deux programmes Eurêka seraient décidés avant la fin de l'été,"
Le Monde, 30 May 1985, p. 3.
84. Interviews 14, 59 and 61.

approached, German Research Minister Riesenhuber stated that about half the German government's contribution to EUREKA projects would go to research on environmental topics. At Hanover the Germans proposed three specific projects: a Europe-wide research communications network, high-performance lasers, and a data bank on regional air pollutants.[85]

In contrast, the French and British strongly urged that projects advance technologies with good commercial prospects in the immediate term. Beyond agreeing on that, they had some differences. The initial French proposal listed technologies similar to those covered by SDI: optoelectronics, new materials, lasers and particle beams, AI, supercomputers, and high-speed microelectronics.[86] Some of these areas clearly have short-term market prospects; with others, like particle beams, the time it would take to market products resulting from the research is not so clear.

The first attempt to define a technical agenda for EUREKA was a set of three extremely general fields of action agreed on by the French and German research and foreign ministers in advance of the Milan Summit. The areas clearly reflected French and German priorities, as they included technologies geared toward the market, like IT; technologies for solving common problems, like environmental protection and toxicology; and large infrastructure projects, like fast transport and telecommunications.[87]

At the time of the first EUREKA conference, in July 1985, some states advanced more specific ideas. British Foreign Secretary Howe stressed that the projects should have potential world markets. The British published a document titled "EUREKA: One of the Paths to Success," emphasizing three areas: transport, technologies for the home, and factories of the future.[88] Two weeks before the conference France had released its own booklet, "The Technological Renaissance of Europe," prepared by the Centre d'Etudes des Systèmes et des Technologies Avancées (CESTA). The CESTA report named five specific areas, giving each a catchy Euro-nym:

85. David Dickson, "Europe Pushes Ahead Plans for Joint Projects," 152; Rupert Cornwell, "Defining EUREKA's Contours," *Financial Times,* 4 November 1985, p. 2.
86. "Au Conseil des Ministres: le projet 'Eurêka' d'une Europe de la technologie,". *Le Monde,* 18 April 1985, p. 30.
87. *Agence Europe,* 28 June 1985, p. 12.
88. *Agence Europe,* 19 July 1985, pp. 7–9.

Euromatique (large computers, AI, fast ICs)

Eurobot (CAD, robotics, factory automation)

Eurocom (IT, broadband telecommunications)

Eurobio (biomedical and agricultural research)

Euromat (new materials)[89]

Clearly, the French wanted a long menu.

The Hanover Declaration approved by EUREKA ministers in November contained a list of advanced technologies to which projects would "initially relate." Included were "information and tele-communication, robotics, materials, manufacturing, biotechnology, marine technology, lasers, environmental protection and transport technologies."[90] These ten areas continue to form the categories by which EUREKA administrators tabulate the projects.

The Hanover conference also announced the first batch of ten projects. Program backers had been eager to have some initial projects ready to announce at Hanover lest EUREKA come to be seen as all talk and no action. In retrospect one EUREKA official acknowledged that some of the early projects were rushed in order to have something concrete to announce and that consequently some of them were not as strong as they could have been.[91] Others, as Foreign Minister Genscher admitted, predated EUREKA but were pulled in to give flesh to the program.[92] The ten projects dealt with subjects as diverse as computers, lasers for cutting fabric and welding, robots with vision (for manufacturing), kits for diagnosing gonorrhea, and the tracking of airborne pollutants.

The Question of Money   Because EUREKA was not to form part of the EC and would have only a minimal administrative structure, it was not clear how projects would find funding. In convening the first EUREKA conference in Paris, President Mitterrand pledged FFr 1 billion for French firms participating in the program. German Research Minister Riesenhuber said his government would set aside

---

89. David Marsh, "EEC Ministers to Discuss Five Lines of Research under EUREKA," *Financial Times*, 3 July 1985, p. 2; *Agence Europe*, 29 July 1985, p. 5.
90. EUREKA, "Declaration of Principles Relating to EUREKA," 2.
91. Interview 43.
92. Rupert Cornwell, "European Technology Projects Adopted," *Financial Times*, 7 November 1985, p. 1.

DM 300 million in the next year's budget to support EUREKA
projects. Other German officials indicated that the figure would come
in part from existing budgets for research and industry and would
depend on marketable projects being proposed. The British were
opposed to public funding; Howe stressed private financing for
projects from European banks.[93] At any rate, it became clear that
funding would vary from country to country.

Even by the fall Britain was still reluctant to discuss public funding. The official position was that the British government would
count on private financial sources, such as the institutions of the
City, to support British firms working in EUREKA. The British hosted
a meeting in London, with representatives of industry and banking
from all the EUREKA countries, to discuss private-sector funding.
Howe told the meeting that EUREKA should not be paid for by
traditional government subsidies.[94] In October the French EUREKA
coordinator, Yves Sillard, declared that each country would provide
funding for those projects it desired and by the methods it thought
best.[95] The Germans had still not made a firm allocation, despite
Riesenhuber's pledge. In fact, Chancellor Kohl had said that Germany would provide public money for EUREKA projects in some
cases but that the government role was more properly one of har-
monizing standards, opening markets, and providing R&D infrastructure.[96] So by the time of the Hanover conference France was
still the only country to have designated specific funds for EUREKA
projects.[97]

British government reluctance to provide state funding for EU-
REKA projects drew the ire of major U.K. companies that planned
to participate in the program. Spokespersons at GEC, Plessey, and
Thorn-EMI called the government position "untenable" and claimed
that it could put British firms at a disadvantage.[98] In March 1986
the government reversed itself. Pattie, minister for IT, announced

93. David Marsh, "Mitterrand Urges Europe to Support EUREKA Project," *Financial Times*, 18 July 1985, p. 48.

94. Peter Marsh and David Marsh, "EUREKA Research Projects Need FFr 50bn
in Funding," *Financial Times*, 14 October 1985, p. 2; *Agence Europe*, 16 October
1985, p. 15.

95. *Agence Europe*, 19 October 1985, p. 14.

96. *Agence Europe*, 7 November 1985, p. 7.

97. Rupert Cornwell, "Defining EUREKA's Contours," *Financial Times*, 4 November 1985, p. 2.

98. Peter Marsh, "Anger over Lack of State Cash for EUREKA," *Financial Times*,
11 October 1985, p. 8.

that British firms participating in EUREKA would be eligible to apply to the DTI for funding of up to half their share of research costs and up to 25 percent of development costs. The total annual research budget of the DTI was £360 million.[99] A British civil servant responsible for EUREKA said that firms seeking funds for EUREKA applied to the Support for Innovation program and had to compete with all other SFI applicants without any special advantage.[100]

In Germany no new funds were allocated to EUREKA in its first year. Rather, some special allowances were made out of existing allocations. For instance, the Bundestag allowed the BMFT to devote to EUREKA DM 40 million in unspent funds that otherwise would have reverted to the Finance Ministry. After that, there was no specific EUREKA allocation. EUREKA projects had to go through the existing programs, principally the BMFT, though EUREKA participants were also free to apply to the *Länder* governments. A BMFT official claimed that these requirements improved the quality of EUREKA projects because they had to compete against other projects in the same technological domain. If there were a EUREKA fund, projects would bid against others from entirely different sectors, making rigorous assessment impossible. The minister of research did however instruct the BMFT program administrators that if two projects were of equal quality, priority must go to the EUREKA project.[101] In July of 1987 the German government finally designated DM 500 million over eight years for EUREKA projects. By that time a significant number of projects were underway with substantial German participation (German parties were involved in about one-third of the 109 announced projects).[102] It is also interesting to note that in the first year German subsidies for EUREKA projects in the area of environmental protection amounted to 100 percent of project costs.[103]

The French contribution of FFr 1 billion was not as impressive as it seemed, in that there were no new funds for EUREKA. Rather, the EUREKA "budget" came from existing programs within var-

99. Peter Marsh, "Aid for Joining EUREKA Project," *Financial Times,* 14 March 1986, p. 8.
100. Interview 30.
101. Interview 61.
102. David Marsh, "Tough Autumn Ahead for Bonn's Science Don," *Financial Times,* 17 July 1987, p. 14.
103. Christian Deubner, *EUREKA: entre les politiques nationales et l'Europe,* 13.

ious agencies, including the Research, Industry, and Telecommunications Ministries and ANVAR. French companies in EUREKA partnerships apply to the appropriate ministry for funding. The EUREKA coordinator discusses funding with the various ministries and reaches agreement on which agency will support which projects. The average level of support is 30 to 40 percent. French EUREKA officials said that in its first year, 1986, EUREKA took money away from programs that had already been budgeted. According to one highly placed administrator, after the first year the ministries spent more on EUREKA but also used their EUREKA burdens as a lever to receive larger allocations.[104] French spending on EUREKA projects rose from FFr 300 million in 1986 to FFr 400 million in 1987 to a projected FFr 790 million in 1988. The increases came after the election of the Chirac government, which apparently did not cut back the French commitment to EUREKA. In fact, the representative of the Chirac government at the ministerial conference in Madrid (September 1987) argued in favor of government funding for the projects.[105]

In the EUREKA system companies and research institutes that form cross-national consortia and qualify for the EUREKA label seek funding individually from their home governments. The participating governments subsidize to varying degrees their own EUREKA participants. In addition to France and Germany, Belgium and the Netherlands publicly announced specific amounts to be targeted for EUREKA projects.[106] The other countries fund projects without any stated total envelope. The French and Germans have assured funding for EUREKA projects; in Britain participants must go through the normal channels, competing with non-EUREKA proposals. The Commission offered to allocate some EC funds for selected EUREKA projects, but the research ministers were divided on the offer. France, Germany, and Denmark were in favor because many of the sectors were similar to those covered in EC programs like ESPRIT and RACE. Ireland, the United Kingdom, the Netherlands, and the southern countries were hesitant. Even the Commission was split, with Commission President Delors favoring closer

104. Interviews 51 and 59.
105. François Grosrichard, "La moitié des projets de recherche européens impliquent des sociétés françaises," Le Monde, 14 March 1988, p. 17; Tom Burns and David Thomas, "Go-Ahead for 58 More EUREKA Projects," Financial Times, 16 September 1987, p. 2.
106. Agence Europe, 9 June 1986, p. 15.

ties with EUREKA and Industry Commissioner Narjes much more reticent.[107]

EUREKA in Operation

Though companies were initially skeptical about a program that had no plan and no money, in time the number of EUREKA projects began to mount. In fact, by November 1985 the Commission claimed to have a computer list of some 600 enterprises, large and small, that had expressed interest.[108] As we have seen, the first ten projects were announced at the Hanover conference that month. The High Representatives, meeting in January 1986 to work out plans for a Secretariat, agreed to sixteen more projects, which would be formally announced at the ministerial conference in London in June. The London conference announced a total of sixty-two new projects, which would require a total expenditure of over $2 billion.[109] EUREKA ministerial conferences have been held once or twice a year, with a new batch of projects announced each time. Table 9.2 shows the number and approximate value of the projects announced at each conference; the actual number of projects is lower because of attrition.

The funding figures represent the project budgets for their full duration, which sometimes reaches eight (two projects), nine (one project), and ten years (five projects). Figures published by the EUREKA Secretariat in mid-September 1987 (after the Madrid conference) showed that 74 percent of the projects would last between two and five years, and only 14 percent would last more than five years.[110]

EUREKA data revealed the following about the size of the projects as of 1987.[111] As Table 9.3 shows, the majority of projects were small, both in the number of countries involved and in cost.[112] The

107. William Dawkins, "Community Divided over Funding for EUREKA Projects," *Financial Times*, 30 June 1988, p. 3.

108. Richard L. Hudson, "Europe's High-Technology Eureka Project," *Wall Street Journal*, 5 November 1985, p. 36.

109. "EUREKA Program Expands," 42.

110. EUREKA, "EUREKA Projects in Figures as of 15th September 1987."

111. EUREKA Secretariat, "All Announced Projects."

112. This is true for the data available after the Madrid conference. The Copenhagen and Vienna conferences did not change the picture drastically because they both announced a relatively large number of projects (fifty-four and eighty-nine, respectively) with a total value of about 1.3 BECU.

TABLE 9.2.   NUMBER OF ANNOUNCED EUREKA
PROJECTS AND THEIR VALUE
(in MECU)

Conference	Date	Number	Total Value
Paris	July 1985	0	0
Hanover	November 1985	10	427
London	June 1986	62	2,342
Stockholm	December 1986	37	730
Madrid	September 1987	58	715
Copenhagen	June 1988	54	360
Vienna	June 1989	89	1,009
Total		310	5,583

SOURCES: European Research Coordination Agency Secretariat, "List of All An-
nounced Projects" (Brussels, March 1988); and European Research Coordination
Agency Secretariat, "List of Projects" (Brussels, 1987).
   Note: The actual totals are slightly less than shown because of some attrition.
The total number of projects as of the end of 1989 was 291.

TABLE 9.3.   EUREKA PROJECTS BY NUMBER OF
PARTICIPANTS (AS OF 1987) AND BY COST (AS OF
1989), AS A PERCENTAGE

Number of Countries Participating	Percentage of Projects	Cost	Percentage of Projects
2	57	≤10 MECU	57
3	20	>10 but ≤ 20 MECU	17
4	8	>20 but ≤ 40 MECU	11
5+	15	>40 MECU	15

SOURCES: European Research Coordination Agency Secretariat, *EUREKA* (Brussels,
1987); European Research Coordination Agency Secretariat, *1989 Project Progress
Report* (Brussels, 1989), 15.

project with the greatest number of participants aimed at developing a research computer network based on OSI standards. It included all the EUREKA countries and the Commission. Among the
largest budgets were 404 MECU for the development of four-

megabit nonvolatile memory chips and feasibility studies for six-
teen-megabit chips (involving Thomson of France and SGS of Italy)
and 327 MECU for development of a database for computer-as-
sisted software design (involving six countries).

Regarding the participation of SMEs, at the Copenhagen conference in June 1988 it was claimed that about half the firms involved in EUREKA were from that category.[113] The following list
is a sample of the better-known European companies involved in
EUREKA projects.

Aeritalia	GEC	Peugeot
Aerospatiale	Imperial Chemicals Inc.	Philips
ASEA-Brown-Boveri	ICL	Plessey
BASF	Italtel	Renault
Bosch	Matra	Saab
British Aerospace	MBB	SGS-Thomson
Bull	Mercedes	Siemens
CASA	Montedison	Thomson
CGE	Nokia	Volkswagen
Fiat	Olivetti	Volvo

The projects covered the full range of technologies. The following list shows the distribution of projects by technical area as a
percentage of the total (100 percent); the total number of projects
is 291:[114]

IT	24
Biotechnologies and medical technologies	19
Robotics and production automation	18
Environment	9
New materials	9
Transport	7

113. Camille Olsen, "Deux cent quatorze projets ont reçu le label Eurêka," *Le
Monde*, 18 June 1988, p. 27.

114. EUREKA Secretariat, "All Announced Projects"; and EUREKA Secretariat,
1989 Project Progress Report, 2–3.

Communications   5

Energy   5

Lasers   4

The telematics sectors (IT and telecommunications) have always accounted for the largest share of projects, though that share has declined slightly over time.

Many of the announced projects appeared relatively unexciting: leather tanning with aluminum instead of chromium compounds (2.5 MECU), the development of new commercial varieties of sunflower seeds (3.6 MECU), the "European Strategic Cigar Automation" project (0.35 MECU), and "Smart Motors for Domestic Appliances" (10.8 MECU). Other projects were large and promised major technological and industrial benefits. The Prometheus project aimed at developing computerized road traffic systems usable throughout Europe to increase efficiency and safety. It would include, for example, a vehicle guidance system that would allow trucks and cars to take the most efficient route, given traffic and road conditions, in cities and on highways. Thirteen European automakers (not including Ford and General Motors) were cooperating in the project and by mid-1987 were reporting positive results from the initial studies.[115]

Another high-profile project is HDTV, which seeks to develop a European system for signal transmission and reception. The goal is to have the system evolve from the existing family of Mac-packet television standards in Europe. HDTV would double the picture clarity of television (making it comparable in quality to a 35-millimeter motion-picture film). Clearly, the markets will be immense, both on the consumer side (there are some 600 million television sets in use worldwide) and on the programming and transmission side. Companies from ten countries cooperate in the HDTV project, which has a budget of 200 MECU for four years and is being coordinated with HDTV work carried out in the RACE program.[116]

A third project is JESSI: Joint European Semiconductor Silicon, intended to develop the next generation of microelectronics design and manufacturing technologies. Announced at Stockholm JESSI

115. John Griffiths, "Europe Opts for West German Car Guidance," *Financial Times,* 25 June 1987, p. 14.
116. Jane Rippeteau, "Whose Hand Will Be on the Horizontal Hold?" *Financial Times,* 20 April 1988, p. 19.

became a source of disputation. The foundation of the project is the collaboration of Siemens and Philips in their Mega project, though JESSI was to be open to other European semiconductor firms, equipment makers, and users. The conflict arose when Siemens hinted that the French-Italian company SGS-Thomson could participate only in a secondary way. This angered both SGS-Thomson and the French government. Negotiations among SGS-Thomson, Siemens, and Philips ensued, with SGS-Thomson citing its leadership in erasable programmable read-only memory-chip (EPROM) technology and arguing that in order to face Japanese and American competition cooperation had to involve all three firms equally. By September 1988 a deal had been worked out that included SGS-Thomson as a full participant in JESSI, which projected a total budget of about $2 billion.[117]

Of course, it is too early to tell how successful EUREKA will be in producing marketable products. A few projects have been direct follow-ons to ESPRIT research; one is the EUREKA Advanced Software Technology project. It aims to produce software-engineering factories based on the industrial version of the ESPRIT Portable Common Tool Environment prototype. One EUREKA project achieved notable success. European Silicon Structures (ES-2, actually begun before EUREKA but brought into the program) by mid-1987 had won over 20 percent of the market for customer-specific ICs.[118] More generally (and very tentatively), a report in Le Monde claimed that only 5 percent of the projects had not brought concrete commercial results.[119]

The number of countries participating has grown with time. By 1988 Iceland, Hungary, and Yugoslavia had been admitted. The participation rates for the various countries are not surprising, though France is by far the most active, and Spain has shown a strong presence. EUREKA data tabulate both participating countries and

117. Guy de Jonquieres, "Europeans Plan Eight-Year Chip Research Scheme," *Financial Times,* 15 March 1988, p. 1; Guy de Jonquieres, "European Chips Plan Clouded by Siemens, SGS-Thomson Dispute," *Financial Times,* 15 April 1988, p. 1; Jennifer Schenker, "Turf War Roils Europe's Version of Sematech," 43–44; Eric le Boucher, "Querelle européenne pour les composants électroniques du futur," *Le Monde,* 6 April 1988, p. 28; "Bataille autour des 'puces': SGS-Thomson en piste," *Le Monde,* 6 September 1988, p. 46.

118. Alan Cane, "ES2 Moves Bespoke Chips Towards Off-the-Peg Prices," *Financial Times,* 10 June 1987, p. 37.

119. François Grosrichard, "La moitié des projets," *Le Monde,* 14 March 1988, p. 17.

countries expressing an interest in the various projects. The following list reflects only actual project participation as of the end of 1989 (by number of projects):[120]

France	130
Germany	101
Italy	88
Spain	85
United Kingdom	85
Netherlands	71
Sweden	49
Austria	36
Belgium	35
Denmark	35
Norway	35
Switzerland	34
Finland	28
Portugal	19
Greece	16
Ireland	8
CEC	6
Luxembourg	5
Turkey	4
Yugoslavia	4
Iceland	3
Canada	2
United States	2
Hungary	1

CONCLUSION

EUREKA is a species of technological collaboration *sui generis*. Instead of a plan, it is an umbrella covering almost every area that

120. EUREKA Secretariat, "All Announced Projects."

could fit in the category of high technology. Instead of emerging out of the EC framework, like ESPRIT and RACE, it came into being in part as a reaction against the growing EC role in technology policy. The issue was not that the Commission had botched its telematics programs; in fact, officials in London, Paris, and Bonn all praised the Commission's management of ESPRIT. At the same time, national governments did not want too much technology policy-making initiative to shift to Brussels. Governments wanted to reassert themselves, especially in high-tech R&D near the market. The consensus, expressed to me by officials in each of the major capitals, was that the Commission was distinctly unsuited to run programs that aimed at short-term marketable results. It was thus the market orientation of the program that, in the view of national governments, ruled out Commission administration. In that sense EUREKA provides an intergovernmental counterpoint to the previous two cases analyzed in this book.

In my theoretical framework, the desire for close national control (even in a collaborative effort) is not surprising. Commission control of ESPRIT and RACE was acceptable because the programs focused on long-term (precompetitive) R&D. To be sure, ESPRIT and RACE could have been run outside EC structures. But policy leadership came from the Commission, not the governments. With EUREKA the initiative came from France and was supported by other governments concerned with the commercial fruits of high-technology R&D. As it did in ESA and Airbus, France assumed a leadership role in European technology.

That said, what does the EUREKA label mean? The bottom line is that the EUREKA label means whatever national coordinators want it to mean. Each country has its own system for running EUREKA. In France the first EUREKA coordinator was Sillard, director of IFREMER.[121] He had a staff of about ten people working out of the IFREMER offices. The Germans assigned a BMFT official, Michael Széplabi, to oversee EUREKA activities; he had two people working on it. EUREKA in the United Kingdom was overseen by Oakley, the civil servant running Britain's Alvey Program. A section of DTI had about five people working full-time on EU-

121. IFREMER is the Institut Français pour l'Exploitation de la Mer, which is in charge of government programs to encourage commercial uses of ocean resources.

REKA.[122] Once all the relevant national coordinators judge that a project meets EUREKA's vague criteria, it receives the EUREKA label. But the EUREKA label does not in every country mean special funding. Some companies participate without state financing.

In addition, EUREKA's administrative structure is minimal. The Secretariat has no responsibility for preparing a work plan, selecting projects, or overseeing their execution. It fulfills the wishes of the large founding states to avoid any sort of centralized bureaucratic agency. The phrases most frequently used to describe the role of the secretariat are *marriage broker* and *information clearinghouse*. The metaphors in this case are well suited to the reality. The Secretariat's database helps parties find partners, and the Secretariat publicizes proposed projects.

Because EUREKA promoters cannot point to a strategic plan or to advantages in funding, a handful of additional measures has become a prime selling point for the program. These have to do with steps that governments can take to enhance the market opportunities for EUREKA products. Principal among these measures is market unification. In fact, the Hanover Declaration states, "The establishment of a large, homogeneous, dynamic and outward-looking European economic area is essential to the success of EUREKA." The Declaration specifically cites the benefits of completing the EC internal market (the 1992 process), including such measures as establishing common standards, eliminating technical obstacles to trade, and opening up public procurement.[123]

The French and British stressed the importance of market opening from the beginning. President Mitterrand emphasized the necessity of unifying the internal market in his speech opening the first ministerial conference in Paris.[124] An official in the French Research Ministry asserted that "the unified market in 1992 is by far the most important thing the Commission can work for, and EUREKA may have pushed people to see the urgency of accelerating the unified market."[125]

The British were even more vigorous proponents than the French of the important connection between EUREKA and the common

122. Interviews 30, 59 and 61.
123. EUREKA, "Declaration of Principles Relating to EUREKA," III.2–III.4.
124. David Marsh, "Mitterrand Urges Europe to Support EUREKA Project," *Financial Times,* 18 July 1985, p. 48.
125. Interview 51.

market,[126] and the British view became the general one. At the London conference the gathered ministers issued a press release stating that they "emphasized the importance and urgency of action to create a single European market of the size needed to enable European industry to compete effectively in world markets for advanced technology. In this context, they welcomed the initiative of the European Community to accelerate the completion of the internal market."[127] Publicity brochures published by the Secretariat stress the supportive measures for market opening that EUREKA offers to potential participants.[128]

Because EUREKA has no central budget, there is no problem of *juste retour*. Industry must take the initiative in defining projects and arranging partnerships; therefore, state officials can presume that commercial interests have been satisfactorily balanced by the firms involved. States contribute financially only to that share of a project accounted for by their own nationals. This sort of cooperation has been called by some (originally the French) variable geometry. It skirts the problem of fair returns and commercial interests.

In sum, the principal object of the EUREKA enterprise is to put into action the consensus that collaboration is essential to Europe's high-tech bootstrapping. European governments grant public monies to companies for the express purpose of collaborating with firms from other European countries. Encouraging cross-national R&D collaboration has become a part of European national technology policy-making.

The EUREKA case bolsters my argument that both the failure of purely national strategies and adaptation on the part of national policy-makers are necessary for collaboration to emerge. EUREKA would never have been suggested or approved without a generalized policy crisis in European high technology, centering on telematics. The French exercised political leadership, proposing, persuading, and mobilizing a coalition. But the EC was crucial to EUREKA as well: The program was borne along on the cooperative momentum generated by ESPRIT and RACE. Commission programs had given cooperation a good name.

---

126. *Agence Europe,* 19 July 1985, p. 9; Rupert Cornwell, "On the Road Certainly, Destination Unknown," *Financial Times,* 8 November 1985, p. 24.
    127. *Agence Europe,* 2 July 1986, p. 7.
    128. EUREKA Secretariat, *EUREKA,* 3.

TEN

Crisis and Collaboration

Western European telematics policies in 1990 looked different than they did in 1980. At the beginning of the decade states were pursuing national-champion strategies, and European collaboration consisted mainly of bitter memories of the Unidata fiasco. By 1986 three major European R&D programs were underway: ESPRIT, RACE, and EUREKA. The collaborative programs by no means replaced the national ones; rather, governments and industry saw them as necessary complements to individual state efforts. The burden of this study has been to explain how the countries of Western Europe moved from decidedly nationalistic telematics policies to collaboration in only five years.

The case studies demonstrate the limits of structural approaches for explaining European telematics collaboration. Structural realists argue that the weak ally to improve their position vis-à-vis the strong. I do not contend that the international context does not matter; it certainly does. By the same token, the claim that states cooperate because they are small or weak is correct, but obvious. More important, the international setting cannot explain why cooperation sometimes emerges and sometimes does not. Put differently, all co-operators must be weak in some way (they need collaborators), but not all of the weak always cooperate. There is variation in outcomes that structural arguments cannot address. The international setting may cast up challenges for national decision-makers, but it by no means accounts for their responses.

Clearly, the weak have alternatives other than alliances among themselves. Take, for instance, the Group of 77 and the NIEO. For small or weak states in the international political economy, cooperation was not (and never is) the obvious or even the best strategy. Korea, Taiwan, Singapore—the Asian newly industrializing countries in general—have pursued unilateral strategies and have grown by tying their fate as fully as possible to the economic system of the industrialized North. Furthermore, the Group of 77 papered over a multitude of diverse and conflicting interests among its members, showing that "international weakness" hides as much as it reveals. Relative weakness was a common plight of the less developed countries. But that explanation is factually true, not analytically powerful. As important in explaining the Group of 77 and its agenda are the dependency ideas that provided them with a common ideology and the impact of political leaders who mobilized a coalition with diverse interests.

To return to European telematics, the distribution of capabilities in the involved sectors, either at the international level or within the region, cannot explain the cooperative outcomes. To point out that Europe was weak relative to the United States and Japan is a truism, not an explanation. National policy-makers in Europe faced multiple options for responding to weakness; European alliance was only one of them. Collaboration need not have been the response at all; states could have opted for revamped, reformed unilateral strategies. Smaller European clubs (like Airbus) were clearly possible in IT. The rise of the Japanese, in fact, gave new freedom of maneuver to Europeans. The possible alliance combinations were increased (and sometimes pursued), with the added opportunity of playing the Japanese against the Americans in order to strike the best deal. In short, there is no analytical path from structural weakness to any outcome in particular. The most that broad structural theories can tell us is that Europeans were likely to respond to relative weakness somehow; but that is not an illuminating insight.

Analysis of the case studies likewise weighs against the realist notion that the dominant concern for states contemplating cooperation is its likely effect on relative capabilities. As Kenneth Waltz puts it, in questions of economic cooperation, "the condition of insecurity—at the very least, the uncertainty of each about the other's future intentions and actions—works against their cooperation." Or, in other words, "a state worries about a division of possible gains that may favor others more than itself."[1] Joseph Grieco elaborates on this theme: "States fear that their partners will achieve relatively greater gains; that, as a result, the partners will surge ahead of them in relative capabilities; and, finally, that their increasingly powerful partners in the present could become all the more formidable foes at some point in the future."[2] Grieco concludes that this concern with "positionality" (as opposed to absolute gains) creates a "relative gains problem" inhibiting cooperation.[3]

The cases examined in this study provide no evidence to support that view. Extensive interviews in Brussels and in national capitals, as well as a thorough review of relevant documents, uncovered no evidence that concern about relative gains played a role in the creation of ESPRIT, RACE, and EUREKA. I did discover, however, that states are concerned about fair returns: They expect a "fair" balance between contributions and rewards. Furthermore, international institutional arrangements are capable of addressing the concerns for fair returns. In the case of Airbus, and in ESA for the large optional programs, shares of contributions and work are fixed explicitly and in advance. In ESPRIT, RACE, and EUREKA, states are able to pursue a fair return over the long run by picking and choosing those parts of the program in which they want to be involved. Furthermore, the key decisions about participation in the telematics cases are made not by governments, but by companies, universities, and laboratories.

In short, relative capabilities are too blunt a tool for analyzing cooperation in the international political economy. It may be that international structure and the relative-gains problem explain much of international security and alliance behavior. But, as this book illustrates, states do not translate every possible interaction into a balance sheet of gains and losses in relative security. Indeed, the telematics cases are strong ones because the technologies involved are so important for industrial and military applications, and yet relative-gains concerns were not a factor.

The theoretical framework elaborated in Chapter 2 approaches the problem of international cooperation from a direction that differs from that of realist theories. Whereas realists tend to explain international phenomena from the top down, starting with inter-

1. Waltz, *Theory of International Politics*, 105–6.
2. Grieco, "Anarchy and the Limits of Cooperation," 499.
3. Ibid., 499.

national structures, my analytical strategy moves through the levels
of analysis from the bottom up. I focus on the processes of cognitive
change that lead decision-makers to adopt collaborative solutions,
the supply of political leadership to organize cooperation, and organizational solutions to the problem of *juste retour*. The next section pulls together findings from all the cases examined in this book
plus two others for comparative scope.

BROAD COMPARISONS

Because state leaders seek to maximize national autonomy, cooperation is rarely their first choice. If the initial preference of states
is always for autonomy, then analysis of international cooperation
must explain the process by which it becomes an acceptable alternative. In other words, theories of international politics must account for the shift in beliefs, perceptions, or preferences that the
choice of cooperation implies. Analysis must address the demand
for cooperation. The question of demand necessarily precedes that
of the structure of strategic games: Leaders of states frequently have
to decide whether they want to enter into a game at all. States pursue unilateral policies not necessarily because that is the equilibrium
outcome in some strategic interaction. It must frequently represent
a decision not to become a player. Analyzing the demand for cooperation, then, means thinking about theories of preference formation and change. It calls for cognitive analysis.

I have suggested that two different kinds of international problems lead to two distinct species of cognitive change. Problems of
scope are those that by their nature require multilateral solutions;
control of environmental pollution is one such issue area. The problem literally crosses borders. Learning occurs as state leaders acquire new information and theories that lead them to recognize
problems of scope for what they are. In recent years the greenhouse
effect and depletion of the ozone layer have been the object of learning: Leaders are acquiring technical information and theories that
provide the bases for cooperative actions to protect the atmosphere,
such as the Montreal Protocols of 1987.[4]

4. Other problems of scope do not require learning as I have defined it. For
instance, coordinating civil air traffic and the handling of international mail are
both problems of scope: They inherently require multilateral attention. In the language of Arthur A. Stein, these are problems of coordination ("Coordination and

Problems of scale pose different kinds of challenges. What needs illuminating in problems of scale is not so much the nature of the problem as the extent of national resources. Large and rich countries can pursue autonomy in more areas than can small or poor ones. The challenge for policy-makers is that frequently the scale of resources that will be required to pursue a goal through unilateral means cannot be accurately predicted. Furthermore, there is often a major element of uncertainty in assessing national resources. Much depends on political capacities—that is, the ability to extract and channel resources: capital, human skills and labor, raw materials. In addition, each policy goal must compete for resources with other objectives. Thus, the total resources that any state can devote to the autonomous pursuit of specific values must be, in many cases, quite unforeseeable and subject to revision.

As a result, with problems of scale states will (given their preference for autonomy) attempt unilateral strategies. In doing so, they will discover the limits of their resources. This process is learning from experience, or adaptation. If national autonomy proves too costly, then state leaders must either revise their ends or alter their means of pursuing them. If the ends are considered too important to abandon, states will initiate a search for alternative approaches. This is the process of policy adaptation described by theorists of organizations.[5] When unilateral strategies become untenable, state leaders search for new policies. While in this adaptive mode, they can be receptive to proposals for cooperation. The case studies in this book display the process of adaptation. States had to discover the limits of autonomy.

In European telematics the most dramatic instance of a state running up against its resource limits in pursuit of autonomy is that of France. Inspired by President Mitterrand's bold vision of a socialist France bursting out of recession ahead of the pack, the government initiated its ambitious plan for the *filière électronique*. The state would channel FFr 140 billion into telematics development, principally through the nationalized champions. By the fall of 1983 the

---

Collaboration: Regimes in an Anarchic World"). Their solution depends on states' agreeing on conventions. To the extent that such cases present an obvious need for cooperation, cognitive processes are less important. States hardly have to learn that there is a problem of scope. By the same token, these are probably the least interesting of cooperation problems.

5. See, for example, Landau, "On the Concept of a Self-Correcting Organization."

clear fact was that France could not afford that scale of effort. And French policy-makers began talking about the need to consider a European approach.[6] In fact, in a sudden change of course, Mitterrand and Fabius became consistent supporters of ESPRIT and RACE, and later they proposed EUREKA.

The *filière électronique* was the final paroxysm of telematics unilateralism in Europe. Each state was rethinking its telematics policies, and European collaboration was a more palatable option than it had ever been before. But this reassessment alone did not make collaboration a reality. It was a moment of choice, when various paths were possible. Political leadership guided the adaptation of states toward cooperation.

Without leadership, the demand for cooperation will remain latent. As I argued in Chapter 2, costs are associated with organizing cooperation. Proposing, persuading, mobilizing, and brokering compromises require an investment of resources: time, energy, personnel, money. Political leaders assume the costs of organizing in return for expected benefits. International-relations scholarship has frequently recognized the need for leadership, but without specifying the general function of leadership. This oversight may be due to the dominant position of realist thinking in the discipline. Leadership is not seen as necessary for paying the costs of organizing but rather as the natural role of powerful states.

Casting the problem as one of meeting the costs of organizing allows for the possibility that hegemonic powers can perform that function. But it can also accommodate analytically the situation in which a small set of countries jointly might exercise leadership. The additional wrinkle that I have proposed is that, in certain circumstances, IOs can act as political entrepreneurs and exercise leadership on behalf of cooperation. I argued that IO leadership is most likely during policy crises, when past national policies have been discredited and state leaders are searching for new approaches. The point is that international political leadership is necessary for cooperation to emerge; IOs can be one source of that leadership. ESPRIT and RACE provide empirical refutation of the realist contention that IOs are irrelevant to the politics of international cooperation.

In addition, success in organizing cooperation depends on finding some means of assuring states a fair return on their contribution

6. See Chapter 7.

TABLE 10.1 CROSS-CASE ANALYSIS OF EUROPEAN
TECHNOLOGICAL COLLABORATION

	Euratom[a]	1970s EEC[b]	Unidata	Airbus	ESA	ESPRIT, RACE	EUREKA
Relative international weakness	Y	Y	Y	Y	Y	Y	Y
Demand for cooperation							
Unilateral strategies failed	N	N	N	Y	Y	Y	Y
States in policy adaptation	Y	N	N	Y	Y	Y	Y
Supply of political leadership							
From a major state or states	N	N	N	Y	P	N	Y
From an IO	Y	Y	N	N	P	Y	N
Arrangements for *juste retour*							
A la carte	N	P	Y	N	Y	Y	Y
Fixed shares	N	N	N	Y	Y	N	N
Cooperation established	N	N	N	Y	Y	Y	Y

Notes: [a]Euratom was one of the original European Communities; its goal was to promote a European atomic-energy industry. It was quickly relegated to marginal tasks because of national industrial rivalries.
[b]This head refers to comprehensive plans for science and technology policy coordination at the EC level, including the Spinelli and Dahrendorf plans. They received minimal implementation.
Y = condition present
N = condition not present
P = condition partially present

to the effort. Prolonged imbalances between contributions and rewards undermine the willingness of states to continue participating. This notion is analogous to the theory of organizations developed by March and Simon. They postulate that if a member of an organization perceives that the contributions she is asked to make outweigh the rewards (or "inducements") she receives, she will be motivated to leave the organization.[7] I have argued that collectivities of persons (in this case, states) operate on a similar logic. The rewards may be nonmaterial, like prestige or moral rectitude. But there must be a balance between contributions and inducements, or, in other words, a fair return.

I further suggested that different kinds of collaborative endeavors lead to different kinds of arrangements to address *juste retour* concerns. If a collaboration aims at producing a single good, shares must be determined at the outset and fixed in agreements resembling contracts. If collaboration embraces a more diffuse objective that can be broken down into smaller bits and pieces over time, then à la carte participation is efficient. It allows states to choose those parts of the overall program that interest them, knowing that in the long run there will be a good chance of achieving a satisfactory share.

Cognitive change, political leadership, and organizational arrangements for *juste retour* therefore constitute the key elements of an analysis of international cooperation. The cases examined in this study all constitute instances of one kind of cognitive change, namely, adaptation. How do the variables play out in the various cases? Table 10.1 summarizes the values of a set of binary variables for each of the European technology programs discussed in this book. One program not addressed in any of the preceding chapters, Euratom, has been added for comparative breadth. Euratom is a collaborative venture for which perceptive, detailed, political studies exist.[8]

What can we learn from the table? First, relative international weakness is not by itself sufficient to account for technological collaboration. Structure does not explain outcomes. To assert that the Europeans collaborate in high technology because they are weaker than the United States and Japan in that domain is merely to an-

7. March and Simon, *Organizations*, pp. 80ff.
8. I rely primarily on Henry R. Nau, *National Politics and International Technology*.

nounce a truism. Furthermore, it implies a logic that does not withstand scrutiny: that weakness causes alliances of the weak. As I
argued before, it is not taxing to find instances in both security and
economic affairs where weakness did not lead even to attempts to
collaborate. Furthermore, when collaboration does arise, there are
other important explanatory variables and no a priori reason for
subordinating them to structural factors. In addition, Table 10.1
cannot tell the whole story. Specifically, in the case of IT, for example, European weakness was obvious twenty years before there
were serious attempts to cooperate.

In contrast, the table displays a close correspondence between
the failure of unilateral approaches and the emergence of collaboration. It is not enough for states to be weak; states must also be
convinced that they cannot satisfactorily redress that weakness
through unilateral means. European states have managed to establish collaboration only in those instances in which unilateral, national strategies had already proven themselves insufficient. The response to failed unilateral strategies was policy adaptation, which
implies openness to new approaches. Thus, in each of the four cases
in which Europe has succeeded in establishing cooperation, unilateral strategies had failed and governments were adapting.[9]

For instance, in civil aircraft, the British Comet would not stay
in the air because of structural defects. The Plowden Report concluded that the only hope for a British civil-aircraft industry lay
with European collaboration.[10] The French Caravelle was a market
flop, and the French also decided that collaboration was the only
viable way for France to participate in the civil-aircraft industry.
For Germany the aircraft industry was politically foreclosed except
through collaboration. Similarly, in the space sector, Great Britain
and France had already concluded they could not afford fully independent national programs. Germany faced the same postwar
constraints in space as it did in aircraft; collaboration was the only
avenue. Even after the collapse of ELDO and ESRO, the ESA emerged
because governments still attached high value to the aerospace industries, and unilateral approaches were not practicable.

9. Of course, another option when unilateral strategies fail is to abandon the
goal. States leaders can decide that if they cannot achieve an objective autonomously, the objective will be abandoned. Again, collaboration is never the sole option. In the cases addressed in this book, governments perceived the high-technology industries in question to be too important to sacrifice.
    10. Newhouse, *Sporty Game*, 123–26.

ESPRIT and RACE most clearly embodied a conscious recognition on the part of governments that unilateral strategies were not viable, and EUREKA grew out of the same policy crisis. By the early 1980s all the European governments had experienced the failure of unilateral, national-champion policies. All of them were in the midst of policy adaptation—searching for new means to bolster their position in telematics. Before unilateral approaches had played themselves out, collaboration could not get off the ground. In fact the 1970s saw the apogee of national-champion strategies. Unidata foundered when the French decided to pursue an independent strategy involving the American company Honeywell. The Commission's initiatives for EC science and technology planning in the 1970s went nowhere because states firmly believed that national means were sufficient.

The atomic-energy sector displays a similar pattern. The United States held a decisive technological lead because of its massive wartime R&D effort. The European countries were eager to develop nuclear energy as the source of electric power for the future; governments were formulating and reformulating nuclear-energy policies. Euratom, one of the original European Communities, had been created to oversee the development of a nuclear-energy industry at the European level. Yet collaboration did not take hold. Governments in France, Germany, and the United Kingdom wanted to promote their national industries and thought that unilateral strategies were viable. Euratom survived as an organization but was relegated to noncommercial tasks such as nuclear-safety experiments and materials testing. States preferred national programs that had not yet been discredited by the experience of failure.[11]

A second notable conclusion justified by the table is that political leadership played a role in every attempt at cooperation except one. The Unidata case differs from the others in that it was organized by private-sector actors—namely, the three computer firms involved. In a strategic sector like computers, however, it is not surprising that the consortium became politicized and ultimately fell victim to government policies for the sector. In all the other cases, political leadership played a role in proposing, persuading, mobilizing, setting agendas, and brokering compromises. Two subsidiary points need to be made here.

11. Nau, *National Politics and International Technology,* chap. 4.

First, IOs have at times played a central, leadership role in organizing cooperation. In some cases political leadership came from one state, as with France in the case of Airbus and EUREKA. But in others IOs have taken the lead. For instance, in space, the European Space Conference of the early 1970s was the midwife for the ESA. Earlier space organizations had foundered. But the new agency emerged from the efforts of the European Space Conference, which combined in a single forum ELDO, ESRO, and CETS. The Conference produced the key compromises that made ESA possible. In addition, some ESRO reforms of the late 1970s provided a model for collaboration that could win broad acceptance.

The EC telematics programs of the 1980s illustrate most dramatically the leadership potential of IOs. In 1982 the Commission, under the guidance of Davignon, had concluded that European telematics collaboration was essential to a healthy future for the European industries specifically and for the region's economy generally. The Commission organized the Roundtable discussions with the industry giants. With the help of company representatives the Commission drafted a proposal for ESPRIT, including the strategic rationale and specific objectives in the work program. In this sense the Commission/industry alliance was moving ahead of governments. The plan presented by the Commission to the governments enjoyed the full backing of the national-champion companies in IT. With the firms lobbying their home governments, the Commission initiative won surprisingly rapid approval. The RACE program followed a similar course, with the added impetus of an ESPRIT program that was widely perceived as successful.

The second subsidiary point to be made concerning IOs is that they succeed in exercising a leadership role only when unilateral strategies have already failed. IO leadership scored its successes only in those cases in which unilateral strategies had been discredited by experience and states were searching for new approaches, namely, ESA, ESPRIT, and RACE. In contrast, cooperation did not emerge despite vigorous IO leadership in those instances (Euratom, 1970s EEC) in which unilateral strategies had not yet been proven inadequate. In other words, if states have reason to believe that past unilateral approaches are unlikely to succeed, they can be receptive to IO leadership.

The EC telematics programs of the 1980s illustrate the role that IOs can play in helping collaboration to gel. By 1980 all the major

EC countries recognized that their national-champion strategies for telematics had not borne fruit. In the ensuing policy crisis each country reconsidered its policies and sought new approaches. Thus, Commission proposals fell on fertile ground. Policy-makers in each national capital were receptive to new ideas for rejuvenating the telematics industries. The end was important, but the traditional means (the national-champion policies of the past fifteen years) had proven inadequate.

The table also summarizes the organizational arrangements employed to allocate the costs and benefits of collaboration. My argument is that international collaboration has to include mechanisms for assuring participating states a satisfactory share of the rewards. For single-purpose collaboration, states must agree in advance on fixed shares. Airbus is the clearest example of that approach. When the goals of collaboration are general or diffuse, the mechanism can be based on self-selection, or what I have called à la carte participation. An ongoing program, with a multitude of contracts and subcontracts, allows states the reasonable expectation that they will obtain a satisfactory share of pieces. On top of that, states (or participating companies) can choose pieces of the work that interest them. Such a system does not eliminate all disputes over fair shares. But it does allow for constant, mid-course corrections in the division of labor and rewards. ESA provides an example of a collaborative endeavor employing both methods of allocation: fixed shares for the compulsory general programs and à la carte for the optional programs.

The conventional wisdom in the study of technological collaboration in Europe has been that programs addressing market-relevant technologies will be scuttled by competitive commercial interests. This study makes the point that commercial interests can be safeguarded by a variety of institutional measures. Airbus is strictly commercial, but the system for allocating shares in each model has worked smoothly on the whole. EUREKA explicitly addresses commercial technologies but functions entirely à la carte. Even ESPRIT and RACE, both precompetitive, focus on technologies with immense commercial importance in the medium term (five to ten years). Sharing the work has not produced divisive disputes as yet.

Finally, the case studies of ESPRIT and RACE suggest theoretical implications not explicitly part of my analytical framework. One of the most striking features of both programs is the crucial

role played by transnational industrial coalitions. The role of the
Roundtable companies in those stories calls to mind the study of
transnational relations that appeared briefly on the scene in the early
1970s. The Twelve Roundtable firms quickly reached a consensus
supporting the Commission's goal of EC-level cooperation. The
Commission/industry alliance sidestepped national governments for
the first phases in preparing ESPRIT. In fact, though governments
were aware of the efforts being led from Brussels, they were not
approached until the Commission had a full-fledged program with
solid industry backing to lay before them.

Furthermore, all the participants whom I interviewed—from the
Commission, industry, and national governments—affirmed that
the Roundtable companies played a crucial role in selling the programs to their home governments. As national champions, the Twelve
carried considerable clout with national policy-makers. Indeed,
governments had always relied on the industry giants in formulating telematics policies. The companies possessed crucial technical
and market information, and were constantly consulted as governments prepared policies for the telematics sectors. Though ESPRIT
and RACE hardly constitute the basis for a new theory of transnational actors, they certainly indicate that some of the conclusions
of the previous efforts were on the mark: Transnational alliances
of nonstate actors can at times exercise a powerful influence on
international politics.

The role of the giant European telematics houses in establishing
the collaborative technology programs illustrates, furthermore, a
critical shortcoming in contemporary realist theories of international politics. The fundamental proposition of realist theorizing is
that the international structure—the distribution of capabilities—
explains the greatest part of state behaviors. In other words, state
leaders respond above all to international constraints and incentives
stemming from the distribution of power. The problem is that national decision-makers must also respond to a potent array of domestic political incentives; they must make the political commitments necessary to stay in power. They must therefore seek to satisfy
enough domestic interests to maintain their hold on the government. Foreign economic policies—for example, to increase trade
protection—frequently arise from the demands of domestic interests.

Robert Putnam suggests that state leaders are thus involved in "two-level games."[12] National leaders may perceive greater gains from satisfying domestic constituencies than from meeting the "demands" of the international system when these conflict. In other words, there is no reason to assign a priori analytical precedence to the international incentives facing state decision-makers. The role of the Roundtable companies in ESPRIT and RACE illustrates one way that domestic interests influence the international political choices of state leaders.

PROSPECTS FOR EUROPEAN TELEMATICS COLLABORATION

My concern in this book has been to analyze the political origins of European telematics collaboration. Now that the programs are underway, a different kind of question arises: What difference will collaboration make? Will Europe's telematics industries be increasingly competitive as a result of ESPRIT, RACE, and EUREKA? Although such queries are valid, precise answers must be left to financial planners and astrologers. I suspect that dramatic improvements in the competitiveness of European telematics industries are not in the stars. At any rate, Japanese and American companies do not seem overly nervous about that prospect. But in a way the question of what difference the programs make is far different (and more interesting) than whether European firms reap the dreamed-of rewards. Indeed, the programs could be judged successful even if European industry fails to make up any ground. In other words, years from now, in assessing ESPRIT, RACE and EUREKA, evaluators may conclude that even though the programs did not shrink the technology gap, Europe was better off than it would have been without collaborating.

On a separate tack, the collaborative programs of the 1980s will matter in the long run if they change the way technology policy is made in Europe. The question is, Will collaboration become a stable part of European technology and industrial policy-making? The answer will depend on the assessments of industry and govern-

12. Robert D. Putnam, "Diplomacy and Domestic Politics: The Logic of Two-Level Games." Putnam's notion remains only a heuristic suggestion; he does not lay out a formal model for two-level games.

ments. If companies and policy-makers perceive that the benefits of collaboration outweigh the costs, then collaboration is likely to endure. The benefits, quite apart from market results, include access to innovations, efficient use of scarce scientific and engineering personnel, and links to needed outside expertise. The costs include the overhead involved in managing collaboration and some loss of autonomous control. Of course, European collaboration will never replace national policies or alliances with American and Japanese firms. But if it is perceived to yield some net benefits, the European dimension will assume a secure place in the European telematics-policy repertoire.

In other words, European collaboration will endure to the extent that it becomes institutionalized.[13] Collaboration will be institutionalized when it becomes part of the status quo, part of the normal way of doing things. More specifically, if enterprises come to see European collaboration as an important element of their technology practices and strategies, they will continue to collaborate, whether or not governments continue to support collaboration with special programs and subsidies. To the extent that ESPRIT, RACE, and EUREKA change the way the telematics industries are run, collaboration will persist. Evidence exists that the programs have altered standard practices and expectations. At the end of Chapter 7 I laid out some initial indications that ESPRIT, the oldest of the programs, has produced noticeable changes in the European telematics business. ESPRIT accelerated the creation of organizations committed to promoting European IT standards (SPAG, X-Open). It stimulated industrial consolidation, at least in the case of SGS-Thomson, as well as private-sector collaborative R&D ventures like the ECRC. Perhaps the most important change introduced by ESPRIT (and also the least measurable by social science) is the new sense of a telematics community linking researchers and technologists across the region.

The new sense of community in the European telematics industries may be important in another respect. The chief bottleneck in European telematics R&D is not finances but rather qualified personnel. A British report noted that in the United Kingdom the shortage of skilled researchers constituted a major drag on efforts to pro-

---

13. By "becoming institutionalized" in this context I mean that roles and relationships turn into stable patterns and form the basis for norms and expectations. I refer to institutions in the broadest sense.

mote the telematics industries.[14] In that regard one British industrialist stated that cooperation allowed his firm to reach the critical mass of resources and expertise without which research could not achieve its goals.[15] One of the directors of Siemens, Hermann Franz, noted that cooperation is unavoidable, even for his company, because of the demand for "highly qualified specialists who cannot be found in sufficient number."[16] Another German Roundtable executive remarked that "the decisive input in Europe is manpower; cooperative R&D can economize on R&D manpower."[17] A technology planner at a French Roundtable company explained the benefits of ESPRIT in terms of personnel. He said that his firm had about 130 researchers working on ESPRIT projects; but those projects involved a total of about 500 researchers, including the contributions of partners. Cooperation thus allowed his company to benefit from the results of research skills far broader than those within his firm.[18]

In short the collaborative programs gave birth to a new sense of community and confidence in Europe's telematics industries. The benefits of such perceptions are intangible but important. In fact, by proving that the Community could act decisively in new areas, ESPRIT, RACE, and EUREKA were the first swell of Euro-optimism, which produced (and drew increased vigor from) the 1992 program.[19]

The new feeling of a technology community may produce concrete benefits, however, in allowing companies to make efficient use of scarce research personnel. Knowing where expertise resides in Europe and what projects are being pursued should enable firms to find the skills they need to complement those of their own technologists. Some researchers also suggest that industry/university ties (such as those that are part of ESPRIT, RACE, and EUREKA) and participation in large research consortia may encourage technological innovation.[20] In addition, to the extent that sharing information

14. Terry Dodsworth, "State High-Tech Scheme 'Suffered Shortcomings,' " *Financial Times,* 7 April 1988, p. 8.

15. Interview 5.

16. Quoted in Siemens, "Expanding Europe's Contribution to the World Components Market," 35.

17. Interview 46.

18. Interview 18.

19. See Wayne Sandholtz and John Zysman, "1992: Recasting the European Bargain."

20. See Lee Tom Perry and Kurt W. Sandholtz, "A 'Liberating Form' for Radical Product Innovation," especially 28–29.

and combining personnel resources become part of the standard way of conducting telematics business in Europe, collaboration will be institutionalized.

Finally, the telematics industries are increasingly global in nature. European companies need access to American and Japanese technology and markets—and vice versa. International isolation is not a recipe for success in the rapidly changing telematics sectors. Consequently, the numerous ties between European and American or Japanese companies are certain to continue and probably increase. The extensive collaboration in sixty-four-megabit memory chips announced by IBM and Siemens is just one important example of what promises to be a continuing phenomenon. This development does not negate the important role to be played by European collaboration. Indeed, the contrary is true: European collaboration has been important because it has made European firms attractive partners. Company executives in Europe repeatedly stressed that they could not hope for balanced corporate alliances unless they were perceived as technologically and industrially attractive partners.

IMPLICATIONS FOR FURTHER STUDY

In spite of the insecurity and violence that constantly cast a shadow over the realm of international politics, states do manage to cooperate. Yet our tools for analyzing the phenomenon are crude. The effort to improve them is fruitful in two ways. First, international collaboration still constitutes a puzzle that challenges us to refine our concepts and theories. The second justification for our work reflects a normative commitment. We need to know how states reduce fear, insecurity, and distrust, in a variety of settings, for the purpose of jointly enhancing the well-being of their peoples.

International collaboration, like most political outcomes, is always the endpoint on a chain of contingent events. The path that leads to collaboration is not predetermined; at any choice point options other than the one actually selected are real possibilities to decision-makers at the time. The perceived array of alternatives provides the field over which political choices are exercised. Presidents, prime ministers, and autocrats do not enjoy the benefit of historical hindsight and political theories to explain why one course of action was causally necessary. To decision-makers the alternatives are real and influence their choices. One task of the student of politics is therefore to explain, via detailed historical investigation, why one option was chosen over others. Explanation requires a theoretically informed analysis of history. At the same time we can retain a healthy sense of modesty by remembering that no theory can account for all contingencies and that there may be more than one path to any given endpoint.

That said, there clearly are fundamental areas where our theoretical constructs need considerable refinement. One prime area for research is the decision-making dilemma that might arise when state leaders face political incentives from different sources: domestic politics and international politics. Technology collaboration in Europe implies that important national political objectives (economic growth and employment) may lead governments to collaborate even though they prefer to maximize their autonomy from other countries. Do they make similar decisions in other sectors? Under what circumstances do domestic political incentives motivate state leaders to reduce national autonomy? For instance, in a hypothetical case, if currency fluctuations seriously disrupt domestic economic activity, governments might be motivated to agree to multilateral monetary management, even though they would prefer to go it alone. This proposition—that the utility of domestic political gains may sometimes outweigh the negative utility of reductions in national autonomy—needs to be fleshed out, both in a theory of decision-making under such circumstances and in empirical instances of it.

Another area where additional work would be fruitful is the role of international institutions (including regimes and IOs). Neoliberal theorists have argued that international institutions are functional in that they reduce information and transaction costs for countries seeking agreements. This notion needs to be tested empirically. How, for instance, would one recognize the positive effects of reduced information costs on decision-making? Perhaps more important, we need to specify further the conditions under which IOs can be policy entrepreneurs. This study has taken some initial steps in that direction.

A third broad theme that begs for increased attention is that of policy crisis and adaptation. State leaders clearly modify their policies in light of experience. The present study proffers evidences of policy crisis and adaptation, and shows how an IO and transnational industry actors channeled the adaptive processes of states toward collaboration. Further work needs to open up the process of adaptation. In other words, we need a better understanding than we now have of the political and bureaucratic processes by which governments judge failures in their international economic policies and define the options for responding. Surely different actors propose different explanations of policy failure and propose diverse answers. We need theory to help us understand how domestic and international politics affect which definitions of the problem and which solutions prevail.

Finally, events in Eastern Europe are casting up new challenges to existing politics and institutions. Many wonder what impact German unification will have on the 1992 project and on European integration beyond 1992. So far, the Commission has responded to the revolutions in the East by pressing for accelerated progress on a social charter and on economic and monetary union (EMU). Member governments agreed to two intergovernmental conferences, one on EMU and one on political union, convened in December 1990. Many EC states are receptive to Commission proposals for rapid development of the Community; only Britain seems viscerally opposed. Even Denmark, traditionally somewhat ambivalent about the EC, is expressing an interest in moving integration forward. Why? Clearly, Germany's neighbors would prefer to see a unified and larger Germany firmly tied into the EC. By the same token Germany will for the foreseeable future have immensely important economic interests in a vibrant EC. In 1987, 54.6 percent of German imports came from the EC, and 52.7 percent of German exports went there.[21] The economic importance of the EC for Germany will not soon change.

Substantial room exists, in short, for the kinds of bargains that will make for active German participation in a dynamic Community. As a result, the EC could assume a central role in constructing whatever common European house emerges after the revolutions and travails in Eastern Europe.[22] Interests and policies are being formed a step at a time; outcomes are still up in the air. Coming years will therefore provide new opportunities for studying the roles of cognitive change and IOs in world politics.

21. Statistical Office of the European Communities, *Basic Statistics of the Community*, 267, 269.
22. See Wayne Sandholtz, "The EC and the East: Integration after the Revolutions of 1989."

Bibliography

Full documentation for the interviews cited in the text cannot be included in the list of references. The research for this book was conducted under the auspices of the University of California, Berkeley. Under the terms of a protocol approved by the Committee for the Protection of Human Subjects, a guarantee of anonymity was extended to all persons who agreed to be interviewed. Complete notes for the interviews are in the author's possession.

Aked, N. H., and P. J. Gummett (1976). "Science and Technology in the European Communities: The History of the COST Projects." *Research Policy* 5:270–94.

Anchordoguy, Marie (Summer 1988). "Mastering the Market: Japanese Government Targeting of the Computer Industry." *International Organization* 42:509–43.

Archbold, Pamela, and John Verity (1 June 1985). "A Global Industry: The Datamation 100." *Datamation* 31:38.

Arlandis, Jacques (October 1987). "Le dilemme des quarante fabricants." *Telecoms Magazine* 8:58–65.

Arnold, Erik, and Ken Guy (1986). *Parallel Convergence: National Strategies in Information Technology.* London: Frances Pinter.

Aronson, Jonathan David, and Peter F. Cowhey (1988). *When Countries Talk: International Trade in Telecommunications Services.* Cambridge, MA: Ballinger.

Axelrod, Robert (1984). *The Evolution of Cooperation.* New York: Basic Books.

Axelrod, Robert, and Robert O. Keohane (1986). "Achieving Cooperation under Anarchy." In Kenneth A. Oye, ed., *Cooperation under Anarchy,* 226–54. Princeton, NJ: Princeton University Press.

Baldwin, David (1980). "Interdependence and Power: A Conceptual Analysis." *International Organization* 34:471–506.

Bar, François (1985). "Telecommunications in the United Kingdom." In Michael G. Borrus et al., eds., *Telecommunications Development in Comparative Perspective: Appendix*, 75–92, Berkeley, CA: Berkeley Roundtable on the International Economy.

Bar, François, and Michael G. Borrus (1987). *From Public Access to Private Connections: Network Policy and National Advantage*. Working Paper No. 28. Berkeley, CA: Berkeley Roundtable on the International Economy.

Barna, Becky (25 May 1979). "The Datamation 50." *Datamation* 25:15.

Baur, Hans (January 1990). "Telecommunications and the Unified European Market." *Telecommunications* 24:33–35.

Borrus, Michael G. (1988). *Competing for Control*. Cambridge, MA: Ballinger.

Borrus, Michael G. et al., eds. (1985). *Telecommunications Development in Comparative Perspective: Appendix*. Berkeley, CA: Berkeley Roundtable on the International Economy.

——— (1985). *Telecommunications Development in Comparative Perspective: The New Telecommunications in Europe, Japan and the U.S.* Working Paper No. 14. Berkeley, CA: Berkeley Roundtable on the International Economy.

Branscomb, L. M. (February 1982). "Electronics and Computers: An Overview." *Science* 215:755–60.

Browne, Malcolm W. (24 August 1986). "The Star Wars Spinoff." *New York Times Magazine*.

Brueckner, Leslie, with Michael G. Borrus (1984). *Assessing the Commercial Impact of the VHSIC Program*. Working Paper No. 5. Berkeley, CA: Berkeley Roundtable on the International Economy.

Catania, B. (1986). *The Many Ways towards IBC*. Turin: Centro Studi e Laboratori Telecomunicazioni.

Caty, Gilbert-François, and Herbert Ungerer (December 1984). "Les télécommunications nouvelle frontière de l'Europe." *Futuribles* 83:29–50.

Cogez, Patrick (1985). "Telecommunications in West Germany." In Michael G. Borrus et al., eds., *Telecommunications Development in Comparative Perspective: Appendix*, 45–74. Berkeley, CA: Berkeley Roundtable on the International Economy.

Cohen, Stephen S., and John Zysman (1987). *Manufacturing Matters: The Myth of the Post-industrial Economy*. New York: Basic Books.

Coles, Peter (12 July 1990). "Is There Profit as Well as Pride?" *Nature* 346:140.

Commission of the European Communities (CEC) (June 1988). *Background Information for a "2nd Call for Proposals" for the RACE Programme: General Information*. Brussels.

——— (June 1987). *Background Information for the "Call for a Reserve List of Tenders" for the RACE Programme: General Information*. Brussels.

———— (September 1985). *Background Information for the Call for Tenders for the RACE Definition Phase.* OTR 50. Brussels.

———— (November 1986). *Communication from the Commission to the Council: EUREKA and the European Technology Community.* COM (86) 664. Brussels.

———— (1982). *Communication from the Commission to the Council: On Laying the Foundations for a European Strategic Programme of Research and Development in Information Technology: The Pilot Phase.* COM (82) 486 final/2. Brussels.

———— (May 1984). *Communication from the Commission to the Council on Telecommunications.* COM (84) 277 final. Brussels.

———— (September 1983). *Communication from the Commission to the Council on Telecommunications: Lines of Action.* COM (83) 573 final. Brussels.

———— (May 1985). *Communication from the Commission to the Council on the Status of the Community Telecommunications Policy.* COM (85) 276 final. Brussels.

———— (May 1986). *Communication from the Commission to the Council: The Second Phase of ESPRIT.* COM (86) 269 final. Brussels.

———— (July 1987). *Draft ESPRIT Workprogramme.* Brussels.

———— (no date). *ESPRIT: Advance Notice of 1987 Call for Proposals.* Brussels.

———— (December 1987). *ESPRIT: 1987 Information Package.* Brussels.

———— (1988). *ESPRIT: 1988 Annual Report.* Brussels.

———— (1987). *ESPRIT: The First Phase: Progress and Results.* EUR 10940 EN. Brussels.

———— (1986). *EUREKA and the European Technology Community.* COM (86) 664 final. Brussels.

———— (October 1987). *European Telecoms Fact Sheet 6: The RACE Programme.* Brussels.

———— (June 1983). *Proposal for a Council Decision Adopting the First European Strategic Programme for Research and Development in Information Technologies (ESPRIT).* COM (83) 258 final. Brussels.

———— (March 1985). *Proposal for a Council Decision on a Preparatory Action for a Community Research and Development Programme in the Field of Telecommunications Technologies: R&D in Advanced Communications-Technologies for Europe (RACE), Definition Phase.* COM (85) 113 final. Brussels.

———— (July 1987). *Proposal for a Council Regulation concerning the European Strategic Programme for Research and Development in Information Technologies (ESPRIT).* COM (87) 313 final. Brussels.

———— (October 1986). *Proposal for a Council Regulation on a Community Action in the Field of Telecommunications Technologies: RACE.* COM (86) 547. Brussels.

———— (November 1984). *Proposal for an Action Plan: RACE,* OTR 25, issue 1. Brussels.

———— (June 1987). *RACE: Consolidated Preliminary Report of the RDP Projects.* OTR 89 final. Brussels.

———— (March 1987). *RACE: Expressions of Interest.* OTR 94, rev. 1. Brussels.

———— (March 1988). *RACE '88: The RACE Programme in 1988.* OTR 112. Brussels.

———— (March 1989). *RACE '89.* Brussels.

———— (June 1988). *RACE Workplan '89.* OTR 200. Brussels.

———— (March 1985). *Report of the Commission to the Council on R&D Requirements in Telecommunications Technologies as Contribution to the Preparation of the R&D Programme: RACE.* COM (85) 145 final. Brussels.

———— (1986). *3rd ESPRIT Conference: Building Momentum.* Brussels.

———— (1988). *Towards a Competitive Community-wide Telecommunications Market in 1992—Implementing the Green Paper on the Development of the Common Market for Telecommunications Services and Equipment—State of Discussions and Proposals by the Commission.* COM (88) 48. Brussels.

———— (June 1987). *Towards a Dynamic European Economy: Green Paper on the Development of the Common Market for Telecommunications Services and Equipment.* COM (87) 290 final. Brussels.

———— (1982). *Towards a European Strategic Programme for Research and Development in Information Technologies.* COM (82) 287. Brussels.

————, Directorate General XIII (September 1989). *ESPRIT: The Project Synopses,* vols. 1–8. Brussels.

Coriat, Benjamin (1989). "Régime réglementaire, structure de marché et competitivité d'entreprise." In Organization for Economic Cooperation and Development–Berkeley Roundtable on the International Economy, eds., *Information Networks and Business Strategies.* Berkeley, CA: Berkeley Roundtable on the International Economy.

Council on Economic Priorities (1988). *Star Wars: The Economic Fallout.* Cambridge, MA: Ballinger.

Cox, Robert W., and Harold K. Jacobson (1973). *The Anatomy of Influence.* New Haven, CT: Yale University Press.

Davis, Dwight B. (April 1985). "Assessing the Strategic Computing Initiative." *High Technology* 5:41–49.

de Cazanove, Rene (October 1987). "La quête d'un futur à douze." *Telecoms Magazine* 8:51–54.

"Deliveries Begin of Fuselage Sections for A330/A340 Aircraft" (29 October 1990). *Aviation Week and Space Technology* 133:29.

de Medici, Marino (July/August 1985). "Europe Sees SDI as Two-Edged Sword." *Europe* 248:10–11.

Deubner, Christian (1987). *EUREKA: entre les politiques nationales et l'Europe.* Paris: Centre d'Information et de Recherche sur l'Allemagne Contemporaine.

de Vos, Dirk (1983). *Governments and Microelectronics: The European Experience.* Science Council of Canada Background Study No. 49. Quebec: Minister of Supply and Services.

Dickson, David (January/February 1985). "Ariane Challenges U.S. on Satellites." *Europe* 243:22–23.

——— (16 May 1986). "Europe Plans Its Own Mini Space Station." *Science* 232:816–17.

——— (1986). "Europe Pushes Ahead Plans for Joint Projects." *Science* 233:152.

——— (April 1985). "France Seeks Joint European Research." *Science* 228:694.

——— (3 August 1984). "France's New Technocrats." *Science* 225:486–87.

——— (25 July 1986). "Redesign of Ariane Is Underway." *Science* 233:411–12.

Donner, Herbert (1986). "The OSI World, Seen from SPAG Europe." SPAG Services, Brussels. Mimeo.

Dosi, Giovanni (1981). "Institutions and Markets in High Technology: Government Support for Microelectronics in Europe." In Charles Carter, ed., *Industrial Policy and Innovation,* 182–202. London: Heinemann.

Dosi, Giovanni, Laura D'Andrea Tyson, and John Zysman (1989). "Trade, Technologies, and Development: A Framework for Discussing Japan." In Chalmers Johnson, Laura D'Andrea Tyson, and John Zysman, eds., *Politics and Productivity: How Japan's Development Strategy Works,* 3–38. Cambridge, MA: Ballinger.

Dreyfack, Kenneth (23 October 1980). "France Wants Bigger Piece of Pie." *Electronics* 53:98–100.

Dubarle, Patrick (May 1985). "Space: Beginnings of a New Competitive Industry." *OECD Observer* 134:11–17.

Duchêne, François, and Geoffrey Shepherd, eds. (1987). *Managing Industrial Change in Western Europe.* London: Frances Pinter.

Etzioni, Amitai (1964). *Modern Organizations.* Englewood Cliffs, NJ: Prentice-Hall.

"EURECA Industrial Contract Signed" (November 1985). *Space Policy* 1:429.

"EUREKA Program Expands" (September 1986). *Europe.*

Euroconsult (August 1990). *World Space Industry Survey: Ten Year Outlook, 1989–90 edition.* Executive Summary reprinted in *Space Policy* 6:250–59.

European Research Coordination Agency (EUREKA) (1985). "Declaration of Principles Relating to EUREKA." Brussels. Mimeo.

——— (n.d.). "EUREKA Projects in Figures as of 15th September 1987." Brussels. Mimeo.

———, Secretariat (no date). "All Announced Projects." Brussels. Mimeo.

——— (November 1987). *EUREKA.* Brussels.

——— (March 1988). "List of All Announced Projects." Brussels. Mimeo.

——— (January 1987). "List of Projects." Brussels. Mimeo.

——— (1989). "List of Projects." Brussels. Mimeo.

——— (1989). *1989 Project Progress Report.* Brussels.

European Strategic Programme for Research and Development in Information Technology (ESPRIT) Review Board (October 1985). *Mid-Term Review of ESPRIT.* Brussels.

——— (May 1989). *The Review of ESPRIT, 1984–1988.* Brussels.

European Space Agency (ESA) (1987). *ESA Annual Report, 1986.* Paris.

———, Industrial Policy Committee (February 1987). *Geographical Distribution of Contracts.* ESA/IPC(87)15. Paris.

Flamm, Kenneth (1987). *Targeting the Computer: Government Support and International Competition.* Washington, DC: Brookings Institution.

Forecasting and Assessment in the Field of Science and Technology (FAST) (1984). *Eurofutures: The Challenges of Innovation.* London: Butterworths.

"France: Le Pays le Plus Numérise" (October 1987). *Telecoms Magazine* 8:49.

Freeman, Christopher (1987). *Technology Policy and Economic Performance: Lessons from Japan.* London: Frances Pinter.

Frohlich, Norman, Joe A. Oppenheimer, and Oran R. Young (1971). *Political Leadership and Collective Goods.* Princeton, NJ: Princeton University Press.

"From Bundespost to Telekom" (December 1987). *Telecommunications Policy* 11:407.

Fusfeld, Herbert I., and Carmela S. Haklisch (November/December 1985). "Cooperative R&D for Competitors." *Harvard Business Review* 63:60.

Gallagher, Robert T. (1 October 1984). "Belgians Set Up Advanced IC Lab." *Electronics* 56:18–19.

——— (10 July 1986). "Europeans Are Counting on Unix to Fight IBM." *Electronics* 57:43.

——— (2 June 1982). "France Urged to Reorganize Electronics." *Electronics* 55:105.

——— (8 September 1982). "French Want National as Partner." *Electronics* 55:104–5.

——— (10 March 1986). "Now Belgium Has Its Own Player in the IC Game." *Electronics* 59:20–22.

——— (26 November 1984). "Stern Spells Europe's Future O-S-I, Not I-B-M." *Electronics Week* 57:43.

——— (24 July 1986). "Suddenly, the Rules Change for Europe's Telecom Business." *Electronics* 59:113–14.

Gilpin, Robert (1987). *The Political Economy of International Relations.* Princeton, NJ: Princeton University Press.

——— (1975). *U.S. Power and the Multinational Corporation.* New York: Basic Books.

——— (1981). *War and Change in World Politics.* Princeton, NJ: Princeton University Press.

Gosch, John (17 May 1984). "Germany Spends for Its High-Tech Future." *Electronics* 57:101–3.

"Grande-Bretagne: des 'Points d'acces' " (October 1987). *Telecoms Magazine* 8:49.

Greenberg, D. S. (24 October 1969). "Britain: New Emphasis on Industrial Research." *Science* 166:485.

——— (26 September 1969). "France: Profit Rather Than Prestige Is New Policy for Research." *Science* 165:1334–37.

——— (31 October 1969). "Italy: OECD Report Finally Emerges." *Science* 166:587.

——— (27 March 1970). "Science in Italy: Reform Effort Takes a Sharp Turn Leftward." *Science* 167:1704–6.

Grieco, Joseph M. (Summer 1988). "Anarchy and the Limits of Cooperation: A Realist Critique of the Newest Liberal Institutionalism." *International Organization* 42:485–507.

——— (September 1983). "Technical Knowledge, Rational Exchange, and International Cooperation: The Cases of Airbus Industrie and the Unidata Computer Consortium." Paper presented at the annual meeting of the American Political Science Association, Chicago.

——— (no date). "Toward a Realist Understanding of International Cooperation." Department of Political Science, Duke University.

Haas, Ernst B. (1975). *The Obsolescence of Regional Integration Theory.* Research Series No. 25. Berkeley, CA: Institute of International Studies.

——— (1968). *The Uniting of Europe.* 2d ed. Stanford, CA: Stanford University Press.

——— (1990). *When Knowledge Is Power.* Berkeley: University of California Press.

——— (April 1980). "Why Collaborate? Issue-Linkage and International Regimes." *World Politics* 32:357–405.

——— (1983). "Words Can Hurt You; or, Who Said What to Whom about Regimes." In Stephen D. Krasner, ed., *International Regimes*, 23–59. Ithaca, NY: Cornell University Press.

Haas, Ernst B., Mary Pat Williams, and Don Babai (1977). *Scientists and World Order.* Berkeley: University of California Press.

Haas, Peter M. (Summer 1989). "Do Regimes Matter? Epistemic Communities and Mediterranean Pollution Control." *International Organization* 43:377–403.

Haklisch, Carmela S. (1986). "Technical Alliances in the Semiconductor Industry." Center for Science and Technology Policy, New York University. Mimeo.

Hall, P. H. (1986). "The Theory and Practice of Innovation Policy: An Overview." In P. H. Hall, ed., *Technology, Innovation and Economic Policy*, 1–34. Oxford: Philip Allan.

———, ed. (1986). *Technology, Innovation and Economic Policy.* Oxford: Philip Allan.

Hardin, Russell (1982). *Collective Action*. Baltimore, MD: Johns Hopkins University Press.

Hart, Jeffrey A. (1986). "British Industrial Policy." In Claude E. Barfield and William A. Schambra, eds., *The Politics of Industrial Policy*, 128–60. Washington, DC: American Enterprise Institute.

——— (1988). "The Politics of Global Competition in the Telecommunications Industry." *Information Society* 5:169–201.

——— (1988). "The Teletel/Minitel System in France." *Telematics and Informatics* 5:21–27.

——— (1986). "West German Industrial Policy." In Claude E. Barfield and William A. Schambra, eds., *The Politics of Industrial Policy*, 161–86. Washington, DC: American Enterprise Institute.

Hayward, Keith (1986). *International Collaboration in Civil Aerospace*. London: Frances Pinter.

Herman, Ros (1986). *The European Scientific Community*. Harlow, Essex: Longman.

Hills, Jill (1986). *Deregulating Telecoms: Competition and Control in the United States, Japan and Britain*. London: Frances Pinter.

——— (1984). *Information Technology and Industrial Policy*. London: Croom Helm.

Holsgrove, Terence, and Marion Howard-Healy (1988). *European Telecommunications at the Crossroads*. Report No. 764. Palo Alto, CA: Business Intelligence Program, SRI International.

Horn, Ernst-Jürgen (1987). "West Germany: A Market-Led Process." In François Duchêne and Geoffrey Shepherd, eds., *Managing Industrial Change in Western Europe*, 41–75. London: Frances Pinter.

Howell, Thomas R., William A. Noellert, Janet H. MacLaughlin, and Alan W. Wolff (1988). *The Microelectronics Race*. Boulder, CO: Westview Press.

"ISDN Makes Strides in West Germany" (May 1990). *Telecommunications* 24:22.

Jacquemin, Alexis (1984). "Introduction: Which Policy for Industry?" In Alexis Jacquemin, ed., *European Industry: Public Policy and Corporate Strategy*, 1–10. Oxford: Clarendon Press.

Jacquemin, Alexis, and Bernard Spinoit (1985). *Economic and Legal Aspects of Cooperative Research: A European View*. Working Document No. 16. Brussels: Centre for European Policy Studies.

Jervis, Robert (1976). *Perception and Misperception in International Politics*. Princeton, NJ: Princeton University Press.

Johnson, Chalmers, Laura D'Andrea Tyson, and John Zysman, eds. (1989). *Politics and Productivity: How Japan's Development Strategy Works*. Cambridge, MA: Ballinger.

Joseph, Jonathan (17 March 1986). "How the Japanese Became a Power in Optoelectronics." *Electronics* 59:50–51.

Kelly, Joseph (15 June 1989). "The Datamation 100: Three Markets Shape One Industry." *Datamation* 35:7.

Kelly, Tim (1987). *The British Computer Industry: Crisis and Development*. London: Croom Helm.

Keohane, Robert O. (1984). *After Hegemony*. Princeton, NJ: Princeton University Press.

——— (1983). "The Demand for International Regimes." In Stephen D. Krasner, ed., *International Regimes*, 141–71. Ithaca, NY: Cornell University Press.

Keohane, Robert O., and Joseph S. Nye, Jr. (1975). "International Interdependence and Integration." In Fred I. Greenstein and Nelson W. Polsby, eds., *International Politics*, vol. 8 of *Handbook of Political Science*, 363–414. Reading, MA: Addison-Wesley.

——— (1977). *Power and Interdependence*. Boston: Little, Brown.

——— (Autumn 1987). "*Power and Interdependence* Revisited." *International Organization* 41:725–53.

———, eds. (1972). *Transnational Relations and World Politics*. Cambridge, MA: Harvard University Press.

Kindleberger, Charles (1973). *The World in Depression*. Berkeley: University of California Press.

Krasner, Stephen D. (April 1976). "State Power and the Structure of International Trade." *World Politics* 28:317–47.

——— (1985). *Structural Conflict*. Berkeley: University of California Press.

———, ed. (1983). *International Regimes*. Ithaca, NY: Cornell University Press.

Laboratoire de Recherche en Economie Appliquée, Centre d'Etudes et de Recherches sur l'Entreprise Multinationale (LAREA/CEREM) (1986). *Les stratégies d'accord des groupes de la CEE: Intégration ou éclatement de l'espace industriel européen*. Nanterre: Université de Paris.

Lafferranderie, Gabriel (27 April 1985). "Les modes de coopération dans le domaine des activités spatiales." In *Les entreprises de coopération technique internationale*, 70–96. Paris: La Société Française pour le Droit International.

Lake, David A. (June 1984). "Beneath the Commerce of Nations: A Theory of International Economic Structures." *International Studies Quarterly* 28:143–70.

Landau, Martin (1973). "On the Concept of a Self-Correcting Organization." *Public Administration Review* 6:533–42.

Langereux, Pierre (14 March 1987). "L'ESA aura besoin de plus de 200 milliards F d'ici à l'an 2000." *Air & Cosmos* 1134:25–28.

Lesourne, Jacques (1984). "The Changing Context of Industrial Policy: External and Internal Developments." In Alexis Jacquemin, ed., *European Industry: Public Policy and Corporate Strategy*, 13–38. Oxford: Clarendon Press.

Lindblom, Charles E. (1977). *Politics and Markets*. New York: Basic Books.

——— (1959). "The Science of 'Muddling Through.'" *Public Administration Review* 19:79–88.

Little, A. D. (1983). *European Telecommunications: Strategic Issues and Opportunities for the Decade Ahead, Executive Report*. Brussels.

Mackintosh, Ian (1986). *Sunrise Europe.* Oxford: Basil Blackwell.

Malerba, Franco (1985). *The Semiconductor Business: The Economics of Rapid Growth and Decline.* London: Frances Pinter.

Manuel, Tom (December 1984). "Cautiously Optimistic Tone Set for 5th Generation." *Electronics Week* 57:57–63.

——— (15 December 1981). "West Wary of Japan's Computer Plan." *Electronics* 54:102–4.

March, James G., and Herbert Simon (1958). *Organizations.* New York: Wiley.

McDougall, Walter A. (April 1985). "Space-Age Europe: Gaullism, Euro-Gaullism, and the American Dilemma." *Technology and Culture* 26:179–203.

McKendrick, George G. (December 1987). "The INTUG View on the EEC Green Paper." *Telecommunications Policy* 11:325–29.

Metcalfe, J. S. (1986). "Technological Innovation and the Competitive Process." In P. H. Hall, ed., *Technology, Innovation and Economic Policy,* 35–64. Oxford: Philip Allan.

Moravscik, Andrew (Winter 1991). "Negotiating the Single Act: National Interests and Conventional Statecraft in the European Community." *International Organization* 45:19–56.

Morgan, Kevin, and Douglas Webber (October 1986). "Divergent Paths: Political Strategies for Telecommunications in Britain, France and West Germany." *West European Politics* 9:56–79.

Morisset, Jacques (13 June 1987). "L'Airbus A.320, héraut de l'industrie européenne." *Air & Cosmos* 1147:69–70.

Morris, John (1 June 1986). "France Plans DP Sell-Off." *Datamation* 32:64–68.

Mowery, David C., and Nathan Rosenberg (1982). "The Influence of Market Demand upon Innovation: A Critical Review of Some Recent Empirical Studies." In Nathan Rosenberg, ed., *Inside the Black Box: Technology and Economics,* 193–241. Cambridge: Cambridge University Press.

Nau, Henry R. (Summer 1975). "Collective Responses to R&D Problems in Western Europe: 1955–1958 and 1968–1973." *International Organization* 29:617–53.

——— (1974). *National Politics and International Technology.* Baltimore, MD: Johns Hopkins University Press.

Nelson, Richard R. (1984). *High-Technology Policies: A Five-Nation Comparison.* Washington, DC: American Enterprise Institute.

Newhouse, John (1982). *The Sporty Game.* New York: Knopf.

Nguyen, Godefroy Dang (1986). "Telecommunications: A Challenge to the Old Order." In Margaret Sharp, ed., *Europe and the New Technologies,* 87–133. Ithaca, NY: Cornell University Press.

Nimroody, Rosy, William Hartung, and Paul Grenier (1988). *Star Wars Spinoffs: Blueprint for a High-Tech America?* New York: Council on Economic Priorities.

Nye, Joseph S., Jr. (January 1988). "Neorealism and Neoliberalism." *World Politics* 40:235–51.

Office of Technology Assessment (OTA) (1987). *International Competition in Services.* OTA-ITE-328. Washington, DC: U.S. Government Printing Office.

———— (1983). *International Competitiveness in Electronics.* OTA-ISC-200. Washington, DC: U.S. Congress.

Olson, Mancur, Jr. (1965). *The Logic of Collective Action.* Cambridge, MA: Harvard University Press.

Organization for Economic Cooperation and Development (OECD) (1975). *Changing Priorities for Government R&D.* Paris.

———— (1968). *Gaps in Technology: Electronic Components.* Paris.

———— (1969). *Gaps in Technology: Electronic Computers.* Paris.

———— (1968). *Gaps in Technology: General Report.* Paris.

———— (1986). *Innovation Policy: France.* Paris.

———— (1981). *The Measurement of Scientific and Technical Activities: Proposed Standard Practice for Surveys of Research and Development.* 4th rev. Paris.

———— (1986). *R&D, Invention and Competitiveness.* OECD Science and Technology Indicators No. 2. Paris.

———— (1985). *Science and Technology Policy Outlook, 1985.* Paris.

———— (1985). *The Semiconductor Industry: Trade Related Issues.* Paris.

———— (1983). *Telecommunications: Pressures and Policies for Change.* Paris.

————, Directorate for Science, Technology and Industry (1986). *Technical Co-operation Agreements between Firms: Some Initial Data and Analysis.* DSTI/SPR/86.20, pts. 1 and 2. Paris.

Oye, Kenneth A. (1986). "Explaining Cooperation under Anarchy: Hypotheses and Strategies." In Kenneth A. Oye, ed., *Cooperation under Anarchy,* 1–24. Princeton, NJ: Princeton University Press.

————, ed. (1986). *Cooperation under Anarchy.* Princeton, NJ: Princeton University Press.

Perry, Lee Tom, and Kurt W. Sandholtz (1988). "A 'Liberating Form' for Radical Product Innovation." In Urs E. Gattiker and Laurie Larwood, eds., *Managing Technological Development: Strategic and Human Resources Issues,* 9–31. New York: Walter de Gruyter.

Peterson, Thane, with Amy Borrus (30 June 1986). "Europe's Chipmakers Pull Out of a Long Losing Streak." *Business Week* 2953:22–24.

Pierre, Andrew J., ed. (1987). *A High Technology Gap?* Council on Foreign Relations Europe-America Series No. 6. New York: New York University Press.

Piore, Michael J., and Charles F. Sabel (1984). *The Second Industrial Divide: Possibilities for Prosperity.* New York: Basic Books.

Pisano, Gary, and David J. Teece (1987). *Collaborative Arrangements and Global Technology Strategy: Some Evidence from the Telecommunications Equipment Industry.* International Business Working Paper No.

IB-10. Berkeley: Center for Research in Management, University of California, Berkeley, Business School.

Pryke, Ian (September/October 1983). "Bound for Space." *Europe* 231:16–17.

Putnam, Robert D. (1988). "Diplomacy and Domestic Politics: The Logic of Two-Level Games." *International Organization* 42:427–60.

Ranci, Pippo (1987). "Italy: The Weak State." In François Duchêne and Geoffrey Shepherd, eds., *Managing Industrial Change in Western Europe*, 111–44. London: Frances Pinter.

"RFA: Les recommendations (en substance) de la Commission Witte" (October 1987). *Telecoms Magazine* 8:91.

Rosenberg, Nathan (1982). "How Exogenous Is Science?" In Nathan Rosenberg, ed., *Inside the Black Box: Technology and Economics*. Cambridge: Cambridge University Press.

———, ed. (1982). *Inside the Black Box: Technology and Economics*. Cambridge: Cambridge University Press.

Roth, Paul (1987). *Meaning and Method in the Social Sciences*. Ithaca, NY: Cornell University Press.

Rothwell, Roy (1986). "Reindustrialization, Innovation and Public Policy." In P. H. Hall, ed., *Technology, Innovation and Economic Policy*, 65–83. Oxford: Philip Allan.

Roundtable of European Industrialists (October 1986). *Clearing the Lines: A Users' View on Business Communications in Europe*. Brussels.

Salomon, Jean-Jacques (1973). *Science and Politics*. Trans. Noël Lindsay. London: Macmillan.

Sancton, Thomas A. (3 August 1987). "Airbus Takes Wing." *Time* 130:40.

Sandholtz, Wayne (Summer 1990). "The EC and the East: Integration after the Revolutions of 1989." *Dialogue* 2:12–16.

——— (August/September 1990). "New Europe, New Telecommunications." Paper presented at the annual meeting of the American Political Science Association, San Francisco.

Sandholtz, Wayne, Jay Stowsky, and Steven K. Vogel (1988). "The Dilemmas of Technological Competition in Comparative Perspective: Is It Guns v. Butter?" Berkeley Roundtable on the International Economy. Mimeo.

Sandholtz, Wayne, and John Zysman (October 1989). "1992: Recasting the European Bargain." *World Politics* 42:95–128.

Schenker, Jennifer (28 April 1988). "Turf War Roils Europe's Version of Sematech." *Electronics* 61:43–44.

Schwarz, Michiel (1979). "European Policies on Space Science and Technology 1960–1978." *Research Policy* 8:204–43.

Servan-Schreiber, Jean-Jacques (1968). *The American Challenge*. Trans. Ronald Steel. New York: Atheneum.

Sharp, Margaret, and Claire Shearman (1987). *European Technological Collaboration*. London: Royal Institute of International Affairs.

Shepherd, Geoffrey (1987). "United Kingdom: Resistance to Change." In François Duchêne and Geoffrey Shepherd, eds., *Managing Industrial Change in Western Europe*, 145–77. London: Frances Pinter.

Shifrin, Carole A. (10 September 1990). "Airbus Industrie Expects to Make Profit This Year for First Time." *Aviation Week and Space Technology* 133:67–68.

Siemens (February 1986). "Expanding Europe's Contribution to the World Components Market." *Siemens Review* 53:32–35.

———— (March 1987). "Japan's Challenge—Europe's Response: Interview with Klaus Ziegler." *Siemens Review* 54:16–20.

Simon, Herbert A. (1976). *Administrative Behavior*. 3d ed. New York: Free Press.

Smith, Kevin (31 May 1983). "UK Pursues Fifth-Generation Computer." *Electronics* 56:101–2.

———— (22 September 1983). "Inmos Forced to Get off the Dole." *Electronics* 56:106.

Snidal, Duncan (1986). "The Game *Theory* of International Politics." In Kenneth A. Oye, ed., *Cooperation under Anarchy*, 25–57. Princeton, NJ: Princeton University Press.

———— (Autumn 1985). "The Limits of Hegemonic Stability Theory." *International Organization* 39:581–614.

Solomon, Laurence P. (25 May 1979). "The Top Foreign Contenders." *Datamation* 25:79–81.

Statistical Office of the European Communities (1989). *Basic Statistics of the Community*. Luxembourg.

Stein, Arthur A. (1983). "Coordination and Collaboration: Regimes in an Anarchic World." In Stephen D. Krasner, ed., *International Regimes*, 141–71. Ithaca, NY: Cornell University Press.

Stowsky, Jay (1986). *Beating Our Plowshares into Double-Edged Swords*. Working Paper No. 17. Berkeley, CA: Berkeley Roundtable on the International Economy.

Tate, Paul (15 November 1985). "Picking Up Speed." *Datamation* 31:64–65.

Teece, David J. (December 1986). "Profiting from Technological Innovation." *Research Policy* 15:285–305.

Teece, David J., Gary Pisano, and Michael Russo (1987). *Joint Ventures and Collaborative Arrangements in the Telecommunications Equipment Industry*. International Business Working Paper No. IB-9. Berkeley: Center for Research in Management, University of California, Berkeley, Business School.

Théodule, Marie-Laure (15 September 1988). "La gamme Trans en pleine mutation vers le RNIS." *Electronique Hebdo* 82:39.

Thompson, James (1967). *Organizations in Action*. New York: McGraw-Hill.

Toulmin, Stephen (1961). *Foresight and Understanding*. New York: Harper Torchbooks.

———— (1972). *Human Understanding*. Princeton, NJ: Princeton University Press.

Tyson, Laura D'Andrea (1987). *Creating Advantage: Strategic Policy for National Competitiveness.* Working Paper No. 23. Berkeley, CA: Berkeley Roundtable on the International Economy.

Tyson, Laura D'Andrea, and John Zysman (1983). "American Industry in International Competition." In Laura D'Andrea Tyson and John Zysman, eds., *American Industry in International Competition: Government Policies and Corporate Strategies,* 15–59. Ithaca, NY: Cornell University Press.

———— (1989). "Preface: The Argument Outlined." In Chalmers Johnson, Laura D'Andrea Tyson, and John Zysman, eds., *Politics and Productivity: How Japan's Development Strategy Works,* xiii–xxi. Cambridge, MA: Ballinger.

————, eds. (1983). *American Industry in International Competition: Government Policies and Corporate Strategies.* Ithaca, NY: Cornell University Press.

"UK Fifth-Generation Computer Passes Final Hurdle" (5 May 1983). *Electronics* 56:76.

Ungerer, Hubert, with Nicholas P. Costello (1988). *Telecommunications in Europe.* Brussels: Commission of the European Communities.

van Tulder, Rob, and Gerd Junne (1988). *European Multinationals in Core Technologies.* New York: Wiley.

———— (1984). *European Multinationals in the Telecommunications Industry.* Amsterdam: Universiteit van Amsterdam.

Vedel, Thierry (6–7 February 1987). "La 'déréglementation' des télécommunications en France." Version of a paper delivered at the Colloque sur les "déréglementations," Paris. Mimeo.

Walsh, John (12 May 1967). "France: First the Bomb, Then the 'Plan Calcul.' " *Science* 156:767–70.

———— (14 May 1982). "France Readies New Research Law." *Science* 216:712–14.

———— (5 May 1967). "Some New Targets Defined for French Science Policy." *Science* 156:626–30.

Waltz, Kenneth N. (1979). *Theory of International Politics.* Reading, MA: Addison-Wesley.

Warde, Ibrahim (1985). "French Telecommunications." In Michael G. Borrus et al., eds., *Telecommunications Development in Comparative Perspective: Appendix,* 117–56. Berkeley, CA: Berkeley Roundtable on the International Economy.

Wassenberg, Arthur F. P. (1987). *Strategies and Tactics of European Industrial Policy-Making.* Rotterdam: Rotterdam School of Management.

Weber, Max (1958). *From Max Weber: Essays in Sociology.* C. Wright Mills and H. H. Gerth, eds. New York: Oxford University Press.

Wilke, John (31 March 1986). "Can Europe Untangle Its Telecommunications Mess?" *Business Week* 2939:46–48.

Williams, Roger (1973). *European Technology: The Politics of Collaboration.* New York: Wiley.

Willott, W. B. (1981). "The NEB Involvement in Electronics and Information Technology." In Charles Carter, ed., *Industrial Policy and Innovation*, 203–212. London: Heinemann.

Young, Oran (1989). *International Cooperation: Building Regimes for Natural Resources and the Environment*. Ithaca, NY: Cornell University Press.

Ypsilanti, Dimitri (January 1985). "The Semiconductor Industry." *OECD Observer* 132:14–20.

NEWSLETTERS AND NEWSPAPERS

Agence Europe (Agence Internationale d'Information pour la Presse), 1979–86

Economist, 1977–90

Europolitique, 1983–87

Financial Times (London), 1983–90

International Herald Tribune, 1984–87

La Tribune, 1985–87

Le Matin, 1985

Le Monde, 1983–88

Les Echos, 1983–84

Liberation, 1986

Los Angeles Times, 1985–91

New York Times, 1985–86

Official Journal (of the European Communities), 1981–88

San Francisco Chronicle, 1988

San Jose Mercury News, 1988

Sunday Times, 1982–84

Times Higher Education Supplement, 1984

Wall Street Journal, 1985–89

Washington Post, 1984

Index

Compositor: Impressions, a Division of Edwards Bros., Inc.
       Text: Sabon
    Display: Gill Sans Medium and Sabon

www.ingramcontent.com/pod-product-compliance
Lightning Source LLC
Chambersburg PA
CBHW021109270326
41929CB00009B/802